国外优秀数学著作
原 版 系 列

内诣零流形映射的尼尔森数的阿诺索夫关系

The Anosov Relation for Nielsen Numbers of Maps of Infra-nilmanifolds

● ［比］布拉姆·大·罗克（Bram De Rock） 著

（英文）

哈尔滨工业大学出版社
HARBIN INSTITUTE OF TECHNOLOGY PRESS

黑版贸审字 08－2019－180 号

图书在版编目(CIP)数据

内诣零流形映射的尼尔森数的阿诺索夫关系＝The Anosov Relation for Nielsen Numbers of Maps of Infra-nilmanifolds:英文/(比)布拉姆·大·罗克 (Bram De Rock)著. —哈尔滨:哈尔滨工业大学出版社,2023.1

ISBN 978-7-5767-0609-3

Ⅰ.①内 …　Ⅱ.①布 …　Ⅲ.①不动点定理－英文 Ⅳ.①O189.2

中国版本图书馆 CIP 数据核字(2023)第 030351 号

NEIYI LINGLIUXING YINGSHE DE NIERSENSHU DE ANUO-SUOFU GUANXI

Copyright © VDM Verlag Dr. Müller Aktiengesellschaft & Co. KG

策划编辑　刘培杰　张永芹
责任编辑　关虹玲
封面设计　孙茵艾
出版发行　哈尔滨工业大学出版社
社　　址　哈尔滨市南岗区复华四道街 10 号　邮编 150006
传　　真　0451－86414749
网　　址　http://hitpress.hit.edu.cn
印　　刷　黑龙江艺德印刷有限责任公司
开　　本　886 mm×1 230 mm　1/32　印张 7.875　字数 197 千字
版　　次　2023 年 1 月第 1 版　2023 年 1 月第 1 次印刷
书　　号　ISBN 978-7-5767-0609-3
定　　价　38.00 元

(如因印装质量问题影响阅读,我社负责调换)

Preface

In this book we examine the 'Anosov relation' for (continuous) self-maps $f : M \to M$ of infra-nilmanifolds M. A map f satisfies the Anosov relation if the Nielsen number $N(f)$ and the Lefschetz number $L(f)$ are equal up to sign, or equivalently if $N(f) = |L(f)|$. These numbers find their origin in fixed point theory and they provide information on the fixed points of the map f. More precisely, one of the main objectives of fixed point theory is to calculate $MF(f)$ which is the minimum number of fixed points of all maps homotopic to f. This number however is not readily computable from its definition and therefore the lower bound $N(f)$ is introduced. F. Wecken and B. Jiang showed that under specific conditions, which infra-nilmanifolds satisfy, this lower bound is sharp, i.e. $MF(f) = N(f)$ (see [30],[53]). But again the Nielsen number is not readily computable from its definition and so one can to try to calculate it in another way. One possibility is to use another number associated to a map f, namely the Lefschetz number $L(f)$. In 1985 D. Anosov proved in [1] the following nice theorem.

Theorem 0.1. For any (continuous) self-map $f : M \to M$ of a nilmanifold M we have that $N(f) = |L(f)|$.

This is a very interesting result since $L(f)$ is, in opposite to the previously introduced numbers, computable from its definition. So for self-maps f of nilmanifolds, D. Anosov showed that one can calculate $MF(f)$. In order not to be disrespectful to the authors of [2], we want to note that this result of D. Anosov is actually a generalization of their result on the tori.

Since D. Anosov's theorem is only valid for a specific class of manifolds, a natural question is of course whether this theorem can again be generalized towards other classes of manifolds. In this respect, D. Anosov himself already showed that his theorem does not hold on the Klein bottle. The goal of this book is to verify whether the Anosov theorem can be generalized towards (certain classes of) infra-nilmanifolds or more generally to verify whether the Anosov relation holds for a specific map $f : M \to M$ of an infra-nilmanifold M.

The generalization towards infra-nilmanifolds is quite natural, since nilmanifolds and the Klein bottle are both examples of infra-nilmanifolds. Moreover infra-nilmanifolds are a well studied class of manifolds which, amongst having other properties, can be studied via a nice algebraic description based on almost-crystallographic groups. This description will turn out to be crucial in the proofs of our results and therefore we recall the main properties and definitions of infra-nilmanifolds in a first chapter.

In this chapter we also show that in general infra-nilmanifolds and solvmanifolds are not the same. Note that solvmanifolds are also a natural generalization since the nilmanifolds and the Klein bottle can also be seen as solvmanifolds. Therefore it is important to note that the techniques developed for solvmanifolds, as in e.g. [27], [28], [31], can not be straightforwardly applied to infra-nilmanifolds.

In a second preliminary chapter we recall the definitions and properties of the numbers (associated to a map) mentioned above. We start with a very general description and then we apply these concepts to maps of infra-nilmanifolds. It turns out that these concepts are well understood for infra-nilmanifolds. Moreover K.B. Lee proved in [38] a nice criterion to verify the Anosov relation for a given map of an infra-nilmanifold. This result is essential for this book and is therefore also introduced in the second chapter.

The second part of this book consists of our extensions of the Anosov theorem. There are actually two possible ways of trying to generalize the Anosov theorem. A first (natural) way is to look for classes of manifolds, other than nilmanifolds, for which the relation holds for all continuous maps of the given manifold. For instance E. Keppelmann and C. McCord established this for exponential solvmanifolds ([31]).

In this book we introduce several classes of infra-nilmanifolds for which the Anosov theorem holds. All these classes are defined by means of the holonomy group of an infra-nilmanifold. As we will explain in detail in the first chapter, the fundamental group of an infra-nilmanifold is a torsion-free almost-crystallographic group (called almost-Bieberbach group) and to such groups one associates a uniquely determined finite group called the holonomy group. For instance the holonomy group of the Klein bottle is $\mathbb{Z}_2$.

The first class of infra-nilmanifolds for which we obtain an extension of the Anosov theorem, are the infra-nilmanifolds with an odd order holonomy group. This is examined in Chapter 3. In order to show that the Anosov theorem holds for these manifolds, we introduce the very useful concept of a periodic sequence associated to a map.

The Klein bottle is not an element of this first class of manifolds, however it is an example of an infra-nilmanifold with a cyclic holonomy group, the easiest class of finite groups. In Chapter 5 we take a closer look to this class of infra-nilmanifolds having a cyclic holonomy group. Because of the observations of D. Anosov we already know that the Anosov theorem certainly does not hold for the whole class. Therefore we have to introduce extra conditions in order to be able to generalize the Anosov theorem. To find this extra condition, we consider the holonomy representation, which is a faithful matrix representation of the holonomy group, determined by the given infra-nilmanifold (the almost-Bieberbach group). To be more precise, since the holonomy group is cyclic we can assume that is generated by x_0 and by using the holonomy representation, we can associate a matrix A to x_0 (see again the first chapter for an exact description). Then, in case -1 is not an eigenvalue of A, we are able to generalize the Anosov theorem towards the given infra-nilmanifold.

In this way we obtain a sufficient condition for infra-nilmanifolds with a cyclic holonomy group for the Anosov theorem to hold. Note that the Klein bottle, or any other infra-nilmanifold with $\mathbb{Z}_2$ as its holonomy group, does not satisfy this condition, since in that case -1 is an eigenvalue of the associated matrix.

Moreover, for flat manifolds we are able to show that this sufficient condition is sharp, i.e. it is also a necessary condition. This means that for flat manifolds with a cyclic holonomy group, we obtained a complete picture. Unfortunately this is not the case for infra-nilmanifolds, since we constructed an example of an infra-nilmanifold with $\mathbb{Z}_2$ as holonomy group (and so the above condition is not satisfied) for which the Anosov theorem does hold. In fact, we introduced this example already in Chapter 3 since, as we will argue below, it is already useful in that chapter. This example nicely shows that the validity of the

Anosov theorem is much more delicate for infra-nilmanifolds than for flat manifolds.

A last class of infra-nilmanifolds for which we are able to prove that the Anosov theorem holds, is the class of the flat orientable generalized Hantzsche-Wendt manifolds. This is a very specific, well studied class of infra-nilmanifolds which, as one might expect, generalizes the classical 3-dimensional Hantzsche-Wendt manifold. Namely, these manifolds are orientable n-dimensional flat manifolds having $\mathbb{Z}_2^{n-1}$ as their holonomy group. Note that although the Klein bottle is a 2-dimensional flat manifold with $\mathbb{Z}_2$ as holonomy group, it is not an element of this class, since the Klein bottle is non-orientable.

The reason for studying this class, lies in the fact that it is the complete opposite of the classes we worked with thus far. To explain this, let's consider again the matrices occurring in the holonomy representation. So for any element x_i of the holonomy group we have an associated matrix A_i. In our first class of infra-nilmanifolds, the condition of being of odd order boils down to the fact that for any A_i, -1 is not an eigenvalue. For the second class, we required an analogous condition, but now only for the matrix associated to the generator of the cyclic holonomy group. It's perhaps important to remark here that -1 can be an eigenvalue of the other elements (non generators) of the holonomy group and this makes things much more complicated.

In contrast to this, we have that for the flat orientable generalized Hantzsche-Wendt manifolds of dimension n, any A_i will have -1 as an eigenvalue (and for many elements even of multiplicity $n - 1$). Nevertheless in Chapter 6 we are still able to show that the Anosov theorem does hold for this class of manifolds.

A second different approach to generalize the Anosov theorem is to search for classes of maps on a certain (type of) manifold, for which the theorem holds. For instance, S. Kwasik and K.B. Lee proved in [34] that the Anosov theorem holds for homotopic periodic maps of infra-nilmanifolds and C. McCord extended this to homotopic periodic maps of infra-solvmanifolds ([45]). Other examples are the virtual unipotent maps introduced and studied in [42] by W. Malfait.

We also followed this approach in this book. In Chapter 3, we examine expanding maps of infra-nilmanifolds and we introduce the nowhere expanding maps of infra-nilmanifolds which are, as one may expect, the antipole of the expanding maps. Again, by using the concept of a periodic sequence associated to a map, we are able to show that the Anosov theorem can be generalized for the latter class of maps. For the expanding maps we obtain a necessary and sufficient condition. To be precise, the Anosov relation holds for a given expanding map $f : M \to M$ of an infra-nilmanifold M if and only if M is orientable.

This last result can be used in order to try to show that the Anosov theorem never holds for non-orientable infra-nilmanifolds. However, this trick does not work on any infra-nilmanifold since there are (non-orientable) infra-nilmanifolds which do not admit expanding maps. The example mentioned before is an example of such an infra-nilmanifold and is therefore presented in Chapter 3. Finally it is important to note that every flat and 2-step nilpotent infra-nilmanifold admit expanding maps which implies that the Anosov theorem never holds for such infra-nilmanifolds if they are non-orientable. A result which will appear to be very useful in Part 3.

In Chapter 4 we examine a third well studied class of maps, namely the Anosov diffeomorphisms. It is well known that not every infra-nilmanifold admits such a map. Up till now, only for the flat manifolds a complete description of which of them admitting an Anosov diffeomorphism is known (see [49]). Therefore we concentrate on flat manifolds. In [41], it is proven that if a flat manifold M admits Anosov diffeomorphisms, then its first Betti number $b_1(M)$ satisfies one of the following: $b_1(M) = 0$, $2 \leq b_1(M) \leq n - 2$ or $b_1(M) = n$ (and all situations occur). In the last case M is a torus and so $N(f) = |L(f)|$ for every continuous self-map f of M. For the other cases, we investigate the possibility of constructing a flat manifold M, with prescribed first Betti number, such that on the one hand M admits an Anosov diffeomorphism f with $N(f) \neq |L(f)|$ and on the other hand M also supports an Anosov diffeomorphism g satisfying $N(g) = |L(g)|$. This is almost always possible except for some very restrictive cases. Namely, for primitive flat manifolds M, i.e $b_1(M) = 0$, in dimension 6 we obtain that the Anosov theorem holds for Anosov diffeomorphisms. On the other hand for n-dimensional flat manifolds M with $b_1(M) = n - 2$ we obtain that the Anosov relation never holds for Anosov diffeomorphisms.

In Part 3 of this book we focus on the infra-nilmanifolds in low dimensions, i.e. up to dimension 4. It appears that many of these infra-nilmanifolds are already covered by our results obtained in the second part (or Anosov's original result). However there are still manifolds which are not yet treated, so examining these manifolds is fruitful as a possible source of inspiration for future research.

In order to handle all infra-nilmanifolds up to dimension 4, we need a list of all possible manifolds. For the flat manifolds we use the description of [7] and for the infra-nilmanifolds (with non-abelian universal covering group) we use the classification of [12].

In Chapter 7 we consider the flat manifolds and we start by showing that in dimension 3 we have, by the known theorems from the second part, already a complete picture. In dimension 4 these theorems cover 53 of the 74 manifolds and the remaining flat manifolds are divided into two groups: the ones with $\mathbb{Z}_2 \oplus \mathbb{Z}_2$ as their holonomy group and the ones with a non-abelian holonomy group. The first group is interesting since it resembles a lot the classes of infra-nilmanifolds we already examined. It turns out that this class provides some nice counter examples to possible questions which arise naturally from the results obtained in the second part. For instance we show that the validity of the Anosov theorem is quite subtle and can not completely be determined from the holonomy representation alone.

For the second group, similar results are obtained in case the flat manifolds have D_8 as their holonomy group. But more interestingly, we find that the Anosov theorem always holds for (4-dimensional) flat manifolds with A_4 as their holonomy group. This leads of course to the question whether or not this holds for any flat manifold with A_4 as holonomy group?

For the infra-nilmanifolds, basically the same calculations need to be done, however there are some complications. Firstly, the flat manifolds are determined by Bieberbach groups, which naturally arise as matrix groups. This matrix representation is no longer automatically provided for the almost-Bieberbach groups, which complicates our calculations. Secondly, the fact that we work with non-abelian universal covering groups puts extra restrictions on the (matrices describing the) endomorphisms, and hence also the affine endomorphisms of the group. These extra restrictions appear to have important consequences for the validity of the Anosov theorem. For instance, we can use them to show that the Anosov theorem always holds for 4-dimensional infra-nilmanifolds with a non-abelian holonomy group. This is a rather surprising result when we compare this to the corresponding situation for flat manifolds.

This leads of course to the question how the structure of the universal covering group influences the validity of the Anosov theorem?

In Part 3 we present for each (almost-)Bieberbach group a proof of, or a counter example to, the Anosov relation for this specific (almost-) Bieberbach group (or infra-nilmanifold). As we already argued above, an analysis of the obtained results can hopefully form the basis for future research.

Bram De Rock
Kortrijk, February 2006

Contents

Part I

Preliminaries

Chapter 1

Maps of infra-nilmanifolds: an algebraic description

In this book we examine the Anosov relation for maps of infra-nilmanifolds. Therefore we start by giving basic definitions and properties concerning maps of infra-nilmanifolds. In a second chapter we introduce the Anosov relation.

Infra-nilmanifolds are a large, well studied class of manifolds which can be nicely described algebraically. To be specific, an infra-nilmanifold is covered by a nilpotent Lie group and is completely determined by its fundamental group which is an almost-Bieberbach group. Our results formulated in part two are obtained by using specific properties and classes of almost-Bieberbach groups. Therefore the first two sections of this chapter contain definitions and properties concerning Lie groups and infra-nilmanifolds.

In the third section we introduce a result of K.B. Lee ([38]) which gives, up to homotopy, a nice algebraic description of maps of infra-nilmanifolds. This is very convenient since the Anosov relation is homotopy invariant.

Because of these algebraic properties of maps of infra-nilmanifolds and the manifolds themselves, we are able to translate the 'topological' problem into an 'algebraic' one. This translation is crucial and therefore we illustrate carefully all the related concepts by means of examples. [1]

1.1 Lie groups

We recall some basic facts concerning nilpotent Lie groups and refer to [40], [50] and [52] for a more detailed study.

---

[1] Most of the examples introduced in Part I can also be found in [17].

Throughout this work G is a connected, simply connected, nilpotent Lie group. An affine endomorphism of G is an element (g, φ) of the semigroup $G \rtimes \mathrm{Endo}(G)$ with $g \in G$ the translational part and $\varphi \in \mathrm{Endo}(G)$ (= the semigroup of all continuous endomorphisms of G) the linear part. The product of two affine endomorphisms is given by $(g, \varphi)(h, \mu) = (g \cdot \varphi(h), \varphi\mu)$ and (g, φ) maps an element $x \in G$ to $g \cdot \varphi(x)$. If the linear part φ belongs to $\mathrm{Aut}(G)$, then (g, φ) is an invertible affine transformation of G. We write $\mathrm{Aff}(G) = G \rtimes \mathrm{Aut}(G)$ for the group of invertible affine transformations of G.

Example 1.1. *The best known example of a connected, simply connected, non-abelian Lie group is the Heisenberg group*

$$H = \left\{ \begin{pmatrix} 1 & y & z \\ 0 & 1 & x \\ 0 & 0 & 1 \end{pmatrix} \mid x, y, z \in \mathbb{R} \right\}.$$

For further use, we will use $h(x, y, z)$ to denote the element $\begin{pmatrix} 1 & y & \frac{1}{3}z \\ 0 & 1 & x \\ 0 & 0 & 1 \end{pmatrix}$.

(The reason for introducing a 3 in the upper right corner lies in the use of this example later on). One easily computes that

$$h(x_1, y_1, z_1)h(x_2, y_2, z_2) = h(x_1 + x_2, y_1 + y_2, z_1 + z_2 + 3x_2 y_1).$$

Let us fix the following elements for use throughout this chapter: $a = h(1, 0, 0)$, $b = h(0, 1, 0)$ and $c = h(0, 0, 1)$. The group N generated by the elements a, b, c has a presentation of the form

$$N = \langle a, b, c \mid [b, a] = c^3, \ [c, a] = [c, b] = 1 \rangle.$$

(We use the convention that $[b, a] = b^{-1}a^{-1}ba$.) Obviously the group N consists exactly of all elements $h(x, y, z)$, for which $x, y, z \in \mathbb{Z}$.

For any connected, simply connected nilpotent Lie group G with Lie algebra $\mathfrak{g}$, it is known that the exponential map $\exp : \mathfrak{g} \to G$ is bijective and we denote by log the inverse of exp.

Example 1.2. *The Lie algebra of H is the Lie algebra of matrices of the form*

$$\mathfrak{h} = \left\{ \begin{pmatrix} 0 & y & z \\ 0 & 0 & x \\ 0 & 0 & 0 \end{pmatrix} \mid x, y, z \in \mathbb{R} \right\}.$$

The exponential map is given by

$$\exp : \mathfrak{h} \to H : \begin{pmatrix} 0 & y & z \\ 0 & 0 & x \\ 0 & 0 & 0 \end{pmatrix} \mapsto \begin{pmatrix} 1 & y & z + \frac{xy}{2} \\ 0 & 1 & x \\ 0 & 0 & 1 \end{pmatrix}.$$

Hence

$$\log : H \to \mathfrak{h} : \begin{pmatrix} 1 & y & z \\ 0 & 1 & x \\ 0 & 0 & 1 \end{pmatrix} \mapsto \begin{pmatrix} 0 & y & z - \frac{xy}{2} \\ 0 & 0 & x \\ 0 & 0 & 0 \end{pmatrix}.$$

For later use, we fix the following basis of $\mathfrak{h}$*:*

$$C = \begin{pmatrix} 0 & 0 & \frac{1}{3} \\ 0 & 0 & 0 \\ 0 & 0 & 0 \end{pmatrix} = \log(c), \ B = \begin{pmatrix} 0 & 1 & 0 \\ 0 & 0 & 0 \\ 0 & 0 & 0 \end{pmatrix} = \log(b),$$

$$and \ A = \begin{pmatrix} 0 & 0 & 0 \\ 0 & 0 & 1 \\ 0 & 0 & 0 \end{pmatrix} = \log(a).$$

For any endomorphism φ of the Lie group G to itself there exists an unique endomorphism φ_* of the Lie algebra $\mathfrak{g}$ (namely the differential $d\varphi$ of φ), making the following diagram commutative:

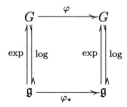

Conversely, every endomorphism φ_* of $\mathfrak{g}$ appears as the differential of an endomorphism of G.

Example 1.3. *Let H and $\mathfrak{h}$ be as before. With respect to the basis C, B and A (in this order!), any endomorphism φ_* is given by a matrix of the form*

$$\begin{pmatrix} k_1 l_2 - k_2 l_1 & l_3 & k_3 \\ 0 & l_2 & k_2 \\ 0 & l_1 & k_1 \end{pmatrix}.$$

This follows from the fact that $3C = [B, A]$ and hence $3\varphi_*(C) = [\varphi_*(B), \varphi_*(A)]$. Conversely, any such a matrix represents an endomorphism of $\mathfrak{g}$. The corresponding endomorphism φ of H satisfies

$\varphi(h(x, y, z))$
$= \exp(\varphi_*(\log(h(x, y, z))))$
$= h(k_1 x + l_1 y, k_2 x + l_2 y,$

$$3k_3 x + 3l_3 y + \frac{3(k_1 x + l_1 y)(k_2 x + l_2 y)}{2} + (k_1 l_2 - k_2 l_1)(z - \frac{3xy}{2})).$$

As one sees, although the map φ_* is linear and thus easy to describe, the corresponding φ is much more complicated. In order to be able to continue presenting examples, we will use a matrix representation of the semigroup $H \rtimes \text{Endo}(H)$ introduced in the previous examples. Given an endomorphism φ of H, let us denote by M_φ the (4×4)-matrix

$$M_\varphi = \begin{pmatrix} P & 0 \\ 0 & 1 \end{pmatrix}$$

where P denotes the (3×3)-matrix, representing φ_* with respect to the basis C, B, A (again in this fixed order). Define the map

$$\psi : H \rtimes \text{Endo}(H) \to M_4(\mathbb{R}) :$$

$$(h(x, y, z), \varphi) \mapsto \begin{pmatrix} 1 & \frac{-3x}{2} & \frac{3y}{2} & \frac{-3xy}{2} + z \\ 0 & 1 & 0 & y \\ 0 & 0 & 1 & x \\ 0 & 0 & 0 & 1 \end{pmatrix} \cdot M_\varphi. \tag{1.1}$$

One easily verifies that ψ defines a faithful representation of the semigroup $H \rtimes \text{Endo}(H)$ into the semigroup $M_4(\mathbb{R})$ (respectively of the group $\text{Aff}(H)$ into the group $\text{Gl}(4, \mathbb{R})$).

Remark 1.4. *An analogous matrix representation can be obtained for any $G \rtimes \text{Endo}(G)$ in case G is two-step nilpotent. (Recall that a group G is said to be k-step nilpotent if the $(k+1)$'th term of the lower central series $\gamma_{k+1}(G) = 1$, where $\gamma_1(G) = G$ and $\gamma_{i+1}(G) = [G, \gamma_i(G)]$. E.g. the Heisenberg group is 2-step nilpotent.) This is proved in [11] for the group $\text{Aff}(G)$, but the details in that paper can easily be adjusted to the case of the semigroup $G \rtimes \text{Endo}(G)$.*

1.2 Infra-nilmanifolds

When we use infra-nilmanifolds we will always refer to the fundamental group which is an almost-Bieberbach group. In this section we give a short introduction to infra-nilmanifolds and some related concepts. We refer to [8] and [12] for more details.

An almost-crystallographic group is a subgroup E of $\text{Aff}(G)$ such that its subgroup of pure translations $N = E \cap G$, is a uniform lattice (by which we mean a discrete and cocompact subgroup) of G and moreover, N is of finite index in E. Therefore the quotient group $F = E/N$ is finite and is called the holonomy group of E. This implies that E fits into the following short exact sequence:

$$1 \to N \to E \to F \to 1$$

Note that the group F is isomorphic to the image of E under the natural projection $\text{Aff}(G) \to \text{Aut}(G)$, and hence F can be viewed as a subgroup of $\text{Aut}(G)$ and of $\text{Aff}(G)$.

Any almost-crystallographic group acts properly discontinuously on (the corresponding) G and the orbit space $E \backslash G$ is compact. Recall that an action of a group E on a locally compact space X is said to be properly discontinuous if for every compact subset C of X, the set $\{\gamma \in E \mid \gamma C \cap C \neq \emptyset\}$ is finite. When E is a torsion-free almost-crystallographic group, it is referred to as an almost-Bieberbach group and the orbit space $M = E \backslash G$ is called an infra-nilmanifold. M is called a nilmanifold in the special case that $E = N$ (and consequently $F = 1$). For (infra-)nilmanifolds E equals the fundamental group $\pi_1(M)$ of the infra-nilmanifold M and we will also talk about F as being the holonomy group of M. When we refer in the sequel to the universal covering group G of an infra-nilmanifold, we actually refer to the above construction.

Any almost-crystallographic group determines a faithful representation $T : F \to \text{Aut}(G)$, which is induced by the natural projection $p : \text{Aff}(G) = G \rtimes \text{Aut}(G) \to \text{Aut}(G)$, and which is referred to as the holonomy representation.

Remark 1.5. *As isomorphic almost-crystallographic subgroups are conjugated inside $\text{Aff}(G)$ (see Theorem 1.10 below or [39]), it follows that the holonomy representation of an almost–crystallographic group is completely determined from the algebraic structure of E up to conjugation by an element of $\text{Aff}(G)$.*

Let $\mathfrak{g}$ denote the Lie algebra of G. By taking differentials, the holonomy representation also induces a faithful representation

$$T_* : F \to \mathrm{Aut}(\mathfrak{g}) : x \mapsto T_*(x) = d(T(x)).$$

This faithful representation will be crucial in the proofs of the results in the second part. As a first application we introduce the following very convenient condition concerning the orientability of an infra-nilmanifold.

Proposition 1.6. *Let M be an infra-nilmanifold with holonomy group F and associated holonomy representation $T : F \to \mathrm{Aut}(G)$. Then*

- *M is orientable $\Leftrightarrow \forall x \in F : \det(T_*(x)) = 1$*
- *M is non-orientable $\Leftrightarrow \exists x \in F : \det(T_*(x)) = -1$*

For more background concerning this proposition see [4, page 211] and [12, page 135].

Example 1.7.
Let φ be the automorphism of H, whose differential φ_* is given by the matrix $\begin{pmatrix} 1 & -\frac{3}{2} & 0 \\ 0 & -1 & 1 \\ 0 & -1 & 0 \end{pmatrix}$. Let $\alpha = (h(0,0,\frac{1}{3}), \varphi) \in \mathrm{Aff}(H)$. Then the group E generated by a, b, c and α has a presentation of the form

$$E = \langle a, b, c, \alpha \mid \begin{array}{lll} [b,a] = c^3 & [c,a] = 1 & [c,b] = 1 \\ \alpha a = b\alpha & \alpha b = a^{-1}b^{-1}\alpha & \alpha c = c\alpha & \alpha^3 = c \end{array} \rangle$$

(this is easily checked using the matrix representation (1.1)). E is an almost-crystallographic group with translation subgroup $N = H \cap E = \langle a, b, c \rangle$ and a holonomy group $F = E/N \cong \mathbb{Z}_3$ of order three (see also [12, page 164, type 13]). We have that

$$T_*(F) = \left\{ I_3, \begin{pmatrix} 1 & -\frac{3}{2} & 0 \\ 0 & -1 & 1 \\ 0 & -1 & 0 \end{pmatrix}, \begin{pmatrix} 1 & -\frac{3}{2} & 0 \\ 0 & -1 & 1 \\ 0 & -1 & 0 \end{pmatrix}^2 = \begin{pmatrix} 1 & 0 & -\frac{3}{2} \\ 0 & 0 & -1 \\ 0 & 1 & -1 \end{pmatrix} \right\}$$

(of course I_n denotes the $(n \times n)$-identity matrix).
As E is torsion-free, it is an almost-Bieberbach group and it determines an infra-nilmanifold $M = E \backslash H$. M is orientable since clearly $\det(T_*(x)) = 1$ for all $x \in F$.

In the special case that G is abelian, i.e. $G = \mathbb{R}^n$, M is a flat manifold. In this case the fundamental group $\pi_1(M)$ is called a torsion-free crystallographic group or Bieberbach group. The fundamental group always fits into an extension

$$1 \to \mathbb{Z}^n \to \pi_1(M) \to F \to 1$$

with F the holonomy group and n the dimension of the manifold. Most of the concepts introduced above are easily adapted but it is worth mentioning that in this case $T_*(x) = T(x)$ since the holonomy representation can also be seen as a map from F to $\mathrm{Gl}(n, \mathbb{Z})$. This simplifies a lot the calculations (and notations), as will be demonstrated later on.

Finally we would like to use the same example once more to show that infra-nilmanifolds and solvmanifolds are different even if at first sight they do not seem so. Although the fundamental group of an infra-nilmanifold with an abelian holonomy group is always solvable (in fact polycyclic), these manifolds do not need to be solvmanifolds in general and so the Nielsen theory on these manifolds cannot be treated by the techniques developed for solvmanifolds (as in e.g. [27], [28], [31]).

Example 1.8. *The almost-Bieberbach group $E = \langle a, b, c, \alpha \rangle$ of Example 1.7 is not the fundamental group of a solvmanifold. Indeed, suppose that E is the fundamental group of a solvmanifold, then it is known that the manifold must admit a Mostow fibering over a torus with a nilmanifold as fibre. On the level of the fundamental group, this implies that there exists a short exact sequence*

$$1 \to \Gamma \to E \to A \to 1 \tag{1.2}$$

where Γ is a finitely generated torsion-free nilpotent group and A is a free abelian group of finite rank. However, it is easy to see that $[E, E]$ is of finite index in E, and therefore, the only free abelian quotient of E is the trivial group. Therefore, there does not exist a normal nilpotent group $\Gamma \subseteq E$, with E/Γ free abelian. This shows that E is not the fundamental group of a solvmanifold.

Remark 1.9. *To be precise: a group E which admits such a short exact sequence as introduced in 1.2 is called a strongly torsion-free S group, with S a solvable Lie group. Moreover E is the fundamental group of a solvmanifold if and only if E is a strongly torsion-free S group. Note that this condition implies that E is solvable but not vice versa. We refer to [44] for more details.*

1.3 Maps of infra-nilmanifolds

In the previous section we saw that infra-nilmanifolds are completely determined by an almost-Bieberbach group. To optimally use this algebraic description we need analogous results for (continuous) maps of infra-nilmanifolds. These are provided by K.B. Lee in [38].

He proved the following theorem concerning almost-crystallographic groups.

Theorem 1.10. *Let* $E, E' \subset \mathrm{Aff}(G)$ *be two almost-crystallographic groups. Then for any homomorphism* $\theta : E \to E'$, *there exists a* $g = (\delta, \mathfrak{D}) \in G \rtimes \mathrm{Endo}(G)$ *such that* $\theta(\alpha) \cdot g = g \cdot \alpha$ *for all* $\alpha \in E$.

Important for us is the following corollary of this theorem (we refer to [38] for a detailed proof).

Corollary 1.11. *Let* $M = E \backslash G$ *be an infra-nilmanifold and* $f : M \to M$ *a continuous map of* M. *Then* f *is homotopic to a map* $h : M \to M$ *induced by an affine endomorphism* $(\delta, \mathfrak{D}) : G \to G$.

We say that $(\delta, \mathfrak{D})$ is a homotopy lift of f. Note that one can find the homotopy lift of a given f, by using Theorem 1.10 for the homomorphism $f_* : \pi_1(M) \to \pi_1(M)$ induced by f. In fact, using this method one can characterize all continuous maps, up to homotopy, of a given infra-nilmanifold M. Indeed, since E is finitely generated, one can construct all possible homomorphisms $\theta : E \to E$ and so, by Theorem 1.10, all possible homotopy lifts $(\delta, \mathfrak{D})$ of f.

Example 1.12. *Let* E *be the almost-Bieberbach group of Example 1.7, then there is a homomorphism* $\theta_1 : E \to E$, *which is determined by the images of the generators as follows:*

$$\theta_1(a) = b^2 c^3, \ \theta_1(b) = a^2 c^3, \ \theta_1(c) = c^{-4}, \ and \ \theta_1(\alpha) = c^{-2}\alpha^2.$$

Using the matrix representation (1.1) it is easy to check that θ_1 *really determines an endomorphism of* E *and that this endomorphism is induced by the affine endomorphism* $(h(0,0,0), \mathfrak{D})$, *where*

$$\mathfrak{D}_* = \begin{pmatrix} -4 & 3 & 3 \\ 0 & 0 & 2 \\ 0 & 2 & 0 \end{pmatrix}.$$

There is a second corollary of this theorem which we also want to point out.

Corollary 1.13. *Let $M = E\backslash G$ be an infra-nilmanifold with holonomy group F and $T : F \to \mathrm{Aut}(G)$ the associated holonomy representation. Suppose $f : M \to M$ is a continuous map of M and $(\delta, \mathfrak{D})$ is a homotopy lift of f. Then*

$$\forall x \in F, \exists y \in F : T_*(y)\mathfrak{D}_* = \mathfrak{D}_*T_*(x).$$

<u>Proof:</u> If $\tilde{x} \in E = \pi_1(M)$ is a pre-image of x, then y can be taken as the natural projection of $f_*(\tilde{x})$, where f_* denotes the morphism induced by f on $\pi_1(M)$.   □

Let us again demonstrate this with an example.

Example 1.14. *Let $M = E\backslash H$ be the infra-nilmanifold introduced in Example 1.7 and suppose that $f_1 : M \to M$ is a continuous map inducing the endomorphism θ_1 on $E = \pi_1(M)$. We know already that $f_* = \theta_1$ is induced by $(1, \mathfrak{D})$ and it is easy to check that*

$$\mathfrak{D}_*\varphi_* = \begin{pmatrix} -4 & 3 & 3 \\ 0 & 0 & 2 \\ 0 & 2 & 0 \end{pmatrix} \begin{pmatrix} 1 & -\frac{3}{2} & 0 \\ 0 & -1 & 1 \\ 0 & -1 & 0 \end{pmatrix} = \begin{pmatrix} 1 & 0 & -\frac{3}{2} \\ 0 & 0 & -1 \\ 0 & 1 & -1 \end{pmatrix} \begin{pmatrix} -4 & 3 & 3 \\ 0 & 0 & 2 \\ 0 & 2 & 0 \end{pmatrix} = \varphi_*^2 \mathfrak{D}_*$$

Finally, since Theorem 1.10 is very crucial for our results, we would like to warn the reader that in the paper of K.B. Lee ([38, Theorem 2.2]) there is a slight mistake in the proof of the theorem. However as we will now argue this has no real influence on the result.

In the beginning of the proof, it is claimed that given a continuous map $f : M = E\backslash G \to M = E\backslash G$, inducing a morphism $f_* : E \to E$ on the fundamental group E of M, one finds that, with $N = E \cap G$,

$$\Lambda = N \cap f_*^{-1} f_* (N \cap f_*^{-1}(N))$$

is a normal finite index subgroup of E, with $f_*(\Lambda) \subseteq \Lambda$. This is however incorrect. In fact, in general one can see that the definition of Λ given above implies that $\Lambda = f_*^{-1}(N) \cap N$, which already indicates that something might be wrong with the above claim.

As an example, let $E = K \times \mathbb{Z}^2$, where K is the fundamental group of the Klein bottle, then

$$E = \langle a, b, c, d \parallel ab = ba^{-1}, \ c, d \ \text{central} \rangle.$$

Of course E is a crystallographic group, with translational part $N = \langle a, b^2, c, d \rangle$. Let $f : M \to M$ be the continuous map on the flat manifold with fundamental group E, inducing the homomorphism $f_* : E \to E$ with

$$f_*(d) = c; \ f_*(c) = b; \ f_*(b) = 1 \text{ and } f_*(a) = 1.$$

Then one easily computes that

- $f_*^{-1}(N) = \{a^k b^l c^{2m} d^n \parallel k, l, m, n \in \mathbb{Z}\}$
- $f_*^{-1}(N) \cap N = \{a^k b^{2l} c^{2m} d^n \parallel k, l, m, n \in \mathbb{Z}\}$
- $f_*(f_*^{-1}(N) \cap N) = \{b^{2m} c^n \parallel m, n \in \mathbb{Z}\}$
- $f_*^{-1}(f_*(f_*^{-1}(N) \cap N)) = \{a^k b^l c^{2m} d^n \parallel k, l, m, n \in \mathbb{Z}\}$
- $\Lambda = f_*^{-1}(f_*(f_*^{-1}(N) \cap N)) \cap N = \{a^k b^{2l} c^{2m} d^n \parallel k, l, m, n \in \mathbb{Z}\}$

So $d \in \Lambda$, but $f_*(d) = c \notin \Lambda$, showing that $f_*(\Lambda) \nsubseteq \Lambda$ as was claimed in [38].

Nevertheless, this is of no real influence for the results of [38]. Indeed, all what is needed is a finite index normal subgroup Λ of E, contained in N and satisfying $f_*(\Lambda) \subseteq \Lambda$. Now, suppose $[E : N] = n$ (which is finite since E is an almost-Bieberbach group) then

$$\Lambda = \langle x^n \parallel x \in E \rangle$$

will be such a group (and this Λ is independent of f_*). Most probably K.B. Lee also noticed this since in [37, Lemma 1.1] he introduced the same group.

Chapter 2

The Anosov relation

In the previous chapter we introduced maps of infra-nilmanifolds. In this chapter we discuss what we exactly want to investigate about these maps, namely the so called Anosov relation.

To a self-map $f : M \to M$ of a manifold M one can associate the Nielsen number $N(f)$ and the Lefschetz number $L(f)$. We then say that the Anosov relation holds for this map if $N(f) = |L(f)|$. In 1985, D. Anosov proved in [1] that this relation holds for any continuous map of a nilmanifold. He also showed that this result does not hold in general for infra-nilmanifolds by constructing a counterexample on the Klein bottle.

As the main topic of this book consists of investigating the Anosov relation for maps of infra-nilmanifolds, we first introduce the necessary general concepts concerning the Nielsen and the Lefschetz number. We also explain why fixed point theorists are interested in these numbers and more specifically in the relation between them. This interest is already demonstrated by several publications inspired by the initial result of D. Anosov: [31], [34], [42], [45],....

After introducing the general setting, we again focus in the second section of this chapter to infra-nilmanifolds. In [38], K.B. Lee proved a nice criterion to decide whether the Anosov relation holds for a map of an infra-nilmanifold.

2.1 Fixed point theory

Let $f : X \to X$ be a map of a compact connected manifold and denote the fixed point set of f by $\text{Fix}(f) = \{x \in X \,|\, f(x) = x\}$. One of the main objectives in fixed point theory is to calculate $MF(f)$, the minimum number of fixed points among all maps homotopic to f:

$$MF(f) = \min\{\#\mathrm{Fix}(g) \,|\, g \simeq f\}.$$

Thus, for instance, $MF(f) = 0$ means that there exists a map g homotopic to f such that $g(x) \neq x$ for all $x \in X$.

The motivation for calculating this number finds its origin in the study of dynamical systems. Note that we can create a dynamical system by iterating $f : X \to X$. In the study of such a dynamical system it is very interesting to know how many fixed points the system has and moreover, how many of these fixed points are persistent under small perturbations. This boils exactly down to calculating $MF(f)$.

In principle, to calculate $MF(f)$ it is necessary to examine the fixed point set of every map homotopic to f. To avoid this, several numbers associated to f are defined to provide information on $MF(f)$.

2.1.1 The Lefschetz number

A first number, and perhaps the best known one, is the Lefschetz number $L(f)$.

Definition 2.1. *Let $f : M \to M$ be a map of a compact connected manifold M. Then*

$$L(f) = \sum_i (-1)^i \, \mathrm{Trace}(f_* : H_i(M, \mathbb{Q}) \to H_i(M, \mathbb{Q})).$$

The Lefschetz number $L(f)$ is a (reasonably) computable invariant of f and it is a homotopy invariant. Unfortunately it does not give a lot of information on the fixed points of f. We only know that if $L(f) \neq 0$, then f has at least one fixed point which can not be removed by a homotopy. In general however $L(f)$ does not give exact information about the number of such points. Of course it is sometimes sufficient to observe that a dynamical system has at least one fixed point and in this case $L(f)$ is very useful.

2.1.2 The Nielsen number

We now introduce the Nielsen number $N(f)$ which is an (almost always sharp) lower bound for $MF(f)$, but unfortunately $N(f)$ is not readily computable from its definition. F. Wecken and B. Jiang proved that this lower bound is sharp for all compact, connected manifolds except for surfaces of negative Euler characteristic ([30],[53]). This is interesting for us since infra-nilmanifolds satisfy these conditions. Since $MF(f) = N(f)$ for self-maps f of infra-nilmanifolds, the calculation of the Nielsen number becomes now a main objective.

The definition of $N(f)$ is based on the concept of essential fixed point classes. Fixed point classes can be introduced in two equivalent ways. A first way uses the universal covering space $(\tilde{M}, p)$ of the manifold M. To be more precise, one can divide all possible lifts $\tilde{f}$ of a given map $f : M \to M$ to the universal covering space $\tilde{M}$ into equivalence classes, called lift classes. Two lifts $\tilde{f}$ and $\tilde{f}'$ are said to be equivalent if they are conjugated by an element of the fundamental group $\pi_1(M)$ (viewed as the group of covering transformations of $(\tilde{M}, p)$). The total number of these lift classes (which is a positive integer or ∞) is called the Reidemeister number $R(f)$ and this number is a homotopy invariant.

Let $\langle \tilde{f} \rangle$ denote the lift class containing a lift $\tilde{f}$. One can show that the projection $p(\text{Fix}(\tilde{f}))$ of the fixed point set of this lift, is independent of the chosen representative of the lift class $\langle \tilde{f} \rangle$. Moreover, $p(\text{Fix}(\tilde{f}))$ is a (possibly empty!) subset of $\text{Fix}(f)$. By considering all possible lifts (in fact lift classes) of f, these subsets $p(\text{Fix}(\tilde{f}))$ divide $\text{Fix}(f)$ into the so called fixed point classes of f.

In this way we have associated to each lift class a fixed point class and therefore there are exactly $R(f)$ fixed point classes. Of course not every fixed point class necessarily contains fixed points. In that case, we say that the lift class is associated to an empty fixed point class. Note that this characterization can lead to many 'different' empty fixed point classes. In fact, for a given map f on a closed manifold M, there can be only finitely many non-empty fixed point classes, while $R(f)$ can be infinite. The advantage of this definition of a fixed point class is the characterization of empty fixed point classes, which is needed later on. A possible disadvantage is that we need a detour via the universal covering space.

A second way of defining the concept of a fixed point class avoids the use of the universal covering space. We partition $\text{Fix}(f)$ into equivalence classes, referred to as fixed point classes, by the relation: $x, y \in \text{Fix}(f)$ are f-equivalent if and only if there is a path w from x to y such that w and fw are (rel. endpoints) homotopic. A disadvantage of this

definition is that is not clear what is meant by an empty fixed point
class, although we need this idea. Indeed, in some occasions fixed point
classes can be removed by homotopy, i.e. a nonempty fixed point class
of f can become an empty fixed point class of a homotopic map g. (Or
equivalently this means that the associated lift class no longer has fixed
points.)

It can be shown, at least for non empty fixed point classes, that these
two definitions are equivalent. See for instance [29] for more details.
Fixed point classes which persist under all homotopies are called essential fixed point classes. Therefore to calculate $MF(f)$ we only have
to focus on the essential fixed point classes and we define:

Definition 2.2. *Let* $f : X \rightarrow X$ *be a map of a compact connected
manifold, then* $N(f) =$ *the number of essential fixed point classes of* f.

The problem is that it is very hard to decide whether a class is essential
or not. If an explicit homotopy can be given that deforms a fixed point
class into an empty class, then by definition this class is inessential.
But, to prove that a fixed point class is essential, we must verify that
no homotopy eliminates this class. Of course this is much harder to
realize.

The solution to this problem is to replace the homotopic-theoretic definition of 'essentialness' by an algebraic approximation. Namely, we
associate (when possible) an integer index to each fixed point class and
define a class to be (algebraic) essential if this index is non-zero. Note
that these two notions of essentialness must of course be the same for
the respective fixed point classes.

For general manifolds it is not easy to define this index. However when
we restrict to compact manifolds this is possible in a satisfactory way.
Satisfactory means for instance that the index must be homotopy invariant and that the sum of the indices of all fixed point classes must
be equal to $L(f)$. This last demand is called the normalization of the
index and $L(f)$ is in this setting also called the total algebraic count of
fixed points. We will limit ourselves here to an informal approach introduced by B. Brown in [6]. For a mathematically rigorous presentation
we refer for instance to [5], [21], [29], [32], ...

Let F be a fixed point class of the map $f : X \rightarrow X$ on a compact
manifold X. Then there is an open subset U of X containing F such
that the closure of U intersects $\text{Fix}(f)$ only in F. For n large enough, we

may embed X in an Euclidian space $\mathbb{R}^n$. Consider then for each point $x \in U \backslash F$ the vector in $\mathbb{R}^n$ from x to $f(x)$. Roughly speaking, if all these vectors point in more or less the same direction, we can modify the definition of f on F to move all those points in the direction indicated by the vectors. In fact in this way we produce a map homotopic to f, and identical to f outside U, that has no fixed points on U. Since F can be eliminated in this way, F is said to be inessential. On the other hand, if the vectors do not all point in somewhat the same direction, the vector field on $U \backslash F$ can be thought of as 'winding around' the set F so that it is not possible to modify f by homotopy in order to eliminate F. Consequently, in this case F is essential. In a rigorous development, the amount of 'winding around' of the vector field is measured by an integer, called the index of F.

2.1.3 The Anosov relation

So, we are interested in the Nielsen number because it gives useful information, but unfortunately it is not readily computable from its definition. For the Lefschetz number however, the opposite holds. Therefore we can try to relate the Nielsen number and the Lefschetz number. The best we can hope for is to show that $N(f) = |L(f)|$ since in this case we can compute an almost always sharp lower bound for $MF(f)$. Recall, that for infra-nilmanifolds this lower bound is always sharp. In this context, D. Anosov proved in 1985 the following theorem ([1]):

Theorem 2.3. *For any (continuous) self-map $f : M \to M$ of a nilmanifold M, we have that $N(f) = |L(f)|$.*

He also showed that the theorem does not hold for the Klein bottle by constructing a counterexample. For the reasons mentioned above, the theorem is very convenient and therefore we try to generalize it towards maps of more general manifolds. Since this is the main topic of this book, we want to explicitly define what we examine.

Definition 2.4. *Let M be a closed manifold.*

- *A map $f : M \to M$ satisfies the Anosov relation if and only if $N(f) = |L(f)|$.*
- *M satisfies the Anosov theorem if and only if every map $f : M \to M$ satisfies the Anosov relation.*

In this work we take off from the result of D. Anosov as it concerns infra-nilmanifolds: nilmanifolds and the Klein bottle are infra-nilmanifolds. So it is quite natural to trying generalizing the Anosov theorem towards other classes of infra-nilmanifolds. We have to mention that in literature one already finds several generalizations of this theorem towards (maps of) solvmanifolds (for instance [27], [28],[31]). This is also a natural approach since nilmanifolds and the Klein bottle can also be seen as solvmanifolds. At the end of the first chapter, we clearly demonstrated that the class of solvmanifolds and the class of infra-nilmanifolds do not coincide and therefore the techniques used for solvmanifolds cannot be applied straightforwardly on infra-nilmanifolds (and vice versa).

Finally we want to note that generalizing the Anosov relation is not the only way to try to obtain information on $N(f)$ (and so on $MF(f)$). The Reidemeister number $R(f)$ is another homotopy invariant associated to f and it is also better computable then $N(f)$. From the definitions above it follows that $R(f)$ is an upper bound for $N(f)$, since it also counts the non-essential fixed point classes. However this upper bound can possibly be very bad since $R(f)$ can be infinite. On the other hand a similar relation to the Anosov relation could exist. For instance, B. Norton-Odenthal showed in [48] that $N(f) = R(f)$ for nilmanifolds and P. Wong did the same for a class of manifolds which are totally different from infra-nilmanifolds ([55]). In this book we did not opt to examine this relation, although this could be a possibility for further research.

2.2 Fixed point theory on infra-nilmanifolds

In this section we want to apply the concepts introduced in the previous section to infra-nilmanifolds. This specific setting gives us convenient tools for our research on the Anosov relation. To start with, we note that in [33, Remark 3.7] it is shown that the index of essential fixed point classes of maps on infra-nilmanifolds is always 1 or −1. Although we will not really need this fact, we shortly recall the ideas behind the proof since they give some valuable intuition for the key results mentioned later on. For more details we refer to [33].

Suppose $f : M \to M$ is a map of an infra-nilmanifold M. By a change up to homotopy we may assume that f is induced by (can be lifted to) an affine endomorphism $(\delta, \mathfrak{D})$ of the universal covering Lie group

G of M. By the argumentation at the end of the previous chapter, we know that there exists in $\pi_1(M)$ a normal subgroup of translations, say K, of finite index such that $f_*(K) \subseteq K$. Actually this implies that there exists a nilmanifold N_K which finitely covers M and such that the map f can not only be lifted to the universal covering space, but also to this nilmanifold, say $\bar{f} : N_K \to N_K$. Of course, one of the possible lifts of f to N_K is the map induced by the affine endomorphism $(\delta, \mathfrak{D})$ on the nilmanifold N_K. The other lifts of f to the nilmanifold N_K are induced by an affine endomorphisms $(\delta', \mathfrak{D}') = \alpha(\delta, \mathfrak{D})$ where $\alpha \in \pi_1(M) \subseteq \text{Aff}(G)$.

In [33] it is shown that calculating the index of an essential fixed point class of f boils down to calculating $\frac{L(\bar{f})}{N(\bar{f})}$ for a specific lift $\bar{f} : N_K \to N_K$. This shows that the index of an essential fixed point class is 1 or -1 since $\bar{f}$ is a map of a nilmanifold and so the Anosov relation holds for $\bar{f}$. Finally in the proof of [33, Theorem 3.5] it is shown that $N(\bar{f}) = |\det(I - \mathfrak{D}'_*)|$.

$$\frac{L(\bar{f})}{N(\bar{f})} = \frac{L(\bar{f})}{|L(\bar{f})|} = \text{sgn}(\det(I_n - \mathfrak{D}'_*)) = \pm 1.$$

We gave this description because it illustrates that one has to calculate 'signs of determinants', which of course can result in different signs for different lifts $\bar{f}$. This is completely formalized in the following theorem of K.B. Lee ([38]):

Theorem 2.5. *Let $f : M \to M$ be a continuous map of an infranilmanifold M and let $T : F \to \text{Aut}(G)$ be the associated holonomy representation. If $(\delta, \mathfrak{D}) \in G \rtimes \text{Endo}(G)$ is a homotopy lift of f, then*

- $N(f) = L(f)$ *if and only if $\forall x \in F : \det(I_n - T_*(x)\mathfrak{D}_*) \geq 0$.*
- $N(f) = -L(f)$ *if and only if $\forall x \in F : \det(I_n - T_*(x)\mathfrak{D}_*) \leq 0$.*

It can be easily verified that $T(x)\mathfrak{D}$ are exactly all possible linear parts of the homotopy lifts of f. So the result above can be interpreted as follows: consider all possible homotopy lifts of f and verify if the associated fixed point classes have the same index. Therefore this result generalizes indeed the concepts introduced above. For a more detailed study we refer to [38].

The above theorem is crucial for this book because it is a very convenient tool to generalize the Anosov theorem. This will be demonstrated elaborately in the next part. Here we already use it to prove the following proposition describing a class of maps on infra-nilmanifolds for which the Anosov theorem always holds. Note that we do not claim that such maps exist on all infra-nilmanifolds.

Proposition 2.6. *Let M be an infra-nilmanifold with holonomy group F and associated holonomy representation $T : F \to \mathrm{Aut}(G)$. Let $f : M \to M$ be a continuous map with homotopy lift $(\delta, \mathfrak{D})$. Suppose that $\forall x \in F,\ x \neq 1 : T_*(x)\mathfrak{D}_* \neq \mathfrak{D}_* T_*(x)$. Then*

$$\forall x \in F : \det(I_n - \mathfrak{D}_*) = \det(I_n - T_*(x)\mathfrak{D}_*)$$

and hence $N(f) = |L(f)|$.

<u>Proof:</u> Let $x \neq 1 \in F$. Since $(\delta, \mathfrak{D})$ is obtained from Theorem 1.10, Corollary 1.13 implies that there exists an $y \in F$ such that $T_*(y)\mathfrak{D}_* = \mathfrak{D}_* T_*(x)$. Because of the condition on T_* and $\mathfrak{D}_*$ we know that $x \neq y$. Then

$$\begin{aligned}
\det(I_n - \mathfrak{D}_*) &= \det(T_*(x) - \mathfrak{D}_* T_*(x))\det(T_*(x^{-1})) \\
&= \det(T_*(x^{-1}))\det(T_*(x) - T_*(y)\mathfrak{D}_*) \\
&= \det(I_n - T_*(x^{-1}y)\mathfrak{D}_*).
\end{aligned}$$

Since $x \neq y$ and T_* is faithful, we have that $T_*(x^{-1}y) \neq I_n$. Moreover, for any other $x' \neq 1 \in F$, with $x \neq x'$ and $T_*(y')\mathfrak{D}_* = \mathfrak{D}_* T_*(x')$, we have that $x^{-1}y \neq x'^{-1}y'$. Indeed, suppose that there exists an $x' \in F$, $x \neq x'$, such that $x^{-1}y = x'^{-1}y'$. Then

$$\begin{aligned}
T_*(x^{-1}y)\mathfrak{D}_* = T_*(x'^{-1}y')\mathfrak{D}_* &\Leftrightarrow T_*(x^{-1})\mathfrak{D}_* T_*(x) = T_*(x'^{-1})\mathfrak{D}_* T_*(x') \\
&\Leftrightarrow \mathfrak{D}_* T_*(xx'^{-1}) = T_*(xx'^{-1})\mathfrak{D}_*.
\end{aligned}$$

The last equality is only satisfied when $xx'^{-1} = 1$ which gives us a contradiction. This proves the proposition because any $x \in F$ determines an unique element $x^{-1}y \in F$, and thus all elements of F are obtained. The last conclusion easily follows from Theorem 2.5.   □

Remark 2.7. *The condition in this proposition may seem strange, but actually it boils down to the fact that for these maps there is only one Reidemeister class on the covering nilmanifold. In that case, one easily verifies that for such maps the Anosov relation holds. See for instance [29].*

Let us demonstrate this proposition.

Example 2.8. *Let $M = E \backslash H$ be the infra-nilmanifold introduced in Example 1.7 and suppose that $f_1 : M \to M$ is a continuous map inducing the endomorphism θ_1 on $E = \pi_1(M)$. We know already that $f_* = \theta_1$ is induced by $(1, \mathfrak{D})$ and it is easy to check that*

$$\varphi_* \mathfrak{D}_* = \begin{pmatrix} 1 & -\frac{3}{2} & 0 \\ 0 & -1 & 1 \\ 0 & -1 & 0 \end{pmatrix} \begin{pmatrix} -4 & 3 & 3 \\ 0 & 0 & 2 \\ 0 & 2 & 0 \end{pmatrix} = \begin{pmatrix} -4 & 3 & 3 \\ 0 & 0 & 2 \\ 0 & 2 & 0 \end{pmatrix} \begin{pmatrix} 1 & 0 & -\frac{3}{2} \\ 0 & 0 & -1 \\ 0 & 1 & -1 \end{pmatrix} = \mathfrak{D}_* \varphi_*^2$$

which implies that the map f (or $\mathfrak{D}_*$) satisfies the criteria of the proposition. Indeed we have that

$$\det(I_3 - \mathfrak{D}_*) = \det(I_3 - \varphi_* \mathfrak{D}_*) = \det(I_3 - \varphi_*^2 \mathfrak{D}_*) = -15.$$

We end this section with an explicit formula to calculate $L(f)$ and $N(f)$ for self-maps of infra-nilmanifolds established in a recent preprint of J.B. Lee and K.B. Lee ([37]).

Theorem 2.9. *Let M be an infra-nilmanifold with holonomy group F and associated holonomy representation $T : F \to \mathrm{Aut}(G)$. Let $f : M \to M$ be a continuous map with homotopy lift $(\delta, \mathfrak{D})$. Then*

$$L(f) = \frac{1}{|F|} \sum_{x \in F} \det(I_n - T_*(x)\mathfrak{D}_*) \qquad \text{and}$$

$$N(f) = \frac{1}{|F|} \sum_{x \in F} |\det(I_n - T_*(x)\mathfrak{D}_*)|.$$

This is of course a nice generalization of Theorem 2.5. Note that this theorem implies that for maps f of infra-nilmanifolds we always have that

$$MF(f) = N(f) \geq |L(f)|.$$

However this lower bound can be very bad as is demonstrated in the following example.

Example 2.10. *Let E be the Bieberbach group generated by*

$$(e_1, I_5), \cdots (e_5, I_5), (a, A) = (\begin{pmatrix} 1/4 \\ 0 \\ 0 \\ 0 \\ 0 \end{pmatrix}, \begin{pmatrix} 1 & 0 & 0 & 0 & 0 \\ 0 & -1 & 0 & 0 & 0 \\ 0 & 0 & -1 & 0 & 0 \\ 0 & 0 & 0 & 0 & -1 \\ 0 & 0 & 0 & 1 & 0 \end{pmatrix})$$

where e_i stands for a column vector of length 5 with 1 in the i-th place and 0 everywhere else ($1 \leq i \leq 5$). It is easily verified that E is indeed a Bieberbach group which fits into the following short exact sequence

$$1 \to \mathbb{Z}^5 \to E \to \mathbb{Z}_4 \to 1$$

So we obtain a 5-dimensional flat manifold $M = E \backslash \mathbb{R}^5$ with holonomy group $\mathbb{Z}_4$. Then take

$$(\delta, \mathfrak{D}) = (\begin{pmatrix} 0 \\ 0 \\ 0 \\ 0 \\ 0 \end{pmatrix}, \begin{pmatrix} k\,0\,0\,0\,0 \\ 0\,1\,1\,0\,0 \\ 0\,2\,1\,0\,0 \\ 0\,0\,0\,1\,0 \\ 0\,0\,0\,0\,1 \end{pmatrix})$$

with $k \in 1 + 4\mathbb{Z}$. The affine endomorphism $(\delta, \mathfrak{D})$ induces a map $f : M \to M$ if

$$(\delta, \mathfrak{D})E(\delta, \mathfrak{D})^{-1} = (\delta, \mathfrak{D})E(-\mathfrak{D}^{-1}\delta, \mathfrak{D}^{-1}) \subseteq E$$

This is indeed the fact since for any i, $1 \leq i \leq 5$, we have that

$$(\delta, \mathfrak{D})(e_i, I_5)(-\mathfrak{D}^{-1}\delta, \mathfrak{D}^{-1}) = (\delta + \mathfrak{D}e_i, \mathfrak{D})(-\mathfrak{D}^{-1}\delta, \mathfrak{D}^{-1})$$
$$= (\mathfrak{D}e_i, I_5) \in E$$

since $\mathfrak{D}e_i \in \mathbb{Z}^5$. Secondly because A and $\mathfrak{D}$ commute we also have

$$(\delta, \mathfrak{D})(a, A)(-\mathfrak{D}^{-1}\delta, \mathfrak{D}^{-1}) = (\delta + \mathfrak{D}a, \mathfrak{D}A)(-\mathfrak{D}^{-1}\delta, \mathfrak{D}^{-1})$$
$$= (\mathfrak{D}a, A)$$
$$= ((\mathfrak{D} - I_5)a, I_5)(a, A) \in E$$

since $(\mathfrak{D} - I_5)a \in \mathbb{Z}^5$. If we then use Theorem 2.9, we obtain

$$L(f) = \frac{1}{4} \sum_{x \in \mathbb{Z}_4} \det(I_5 - T_*(x)\mathfrak{D}_*)$$
$$= \frac{1}{4} \sum_{i=0}^{3} \det(I_5 - A^i\mathfrak{D})$$
$$= \frac{1}{4}(0 + 4(1 - k) + 8(-1 + k) + 4(1 - k))$$
$$= 0$$

and

$$N(f) = \frac{1}{4} \sum_{x \in \mathbb{Z}_4} |\det(I_5 - T_*(x)\mathfrak{D}_*)|$$
$$= \frac{1}{4}(0 + 4|1 - k| + 8| - 1 + k| + 4|1 - k|)$$

If we for instance assume that $k \geq 5$ then $N(f) = 4(k - 1)$. So indeed, $|L(f)| \leq N(f)$ but $N(f)$ can be as large as we want.

Part II

The results

As is explained and motivated in the first part, in the following chapters we examine the Anosov relation for infra-nilmanifolds. Our main purpose is trying to generalize the Anosov theorem and in the case that this turns out to be not possible we provide the necessary counterexamples.

There are two possible ways of trying to generalize the Anosov theorem. Firstly, one can look for classes of manifolds, other than nilmanifolds, for which the relation holds for all continuous maps of the given manifold. This was, for instance, established by E. Keppelmann and C. McCord for exponential solvmanifolds ([31]). We try to do this for classes of infra-nilmanifolds. In Chapter 3 we examine for example infra-nilmanifolds with odd order holonomy group, in Chapter 5 infra-nilmanifolds with cyclic holonomy group and in Chapter 6 flat generalized Hantzsche-Wendt manifolds.

Secondly, one can search for classes of maps on a certain (type of) manifold for which the relation holds. For instance, S. Kwasik and K.B. Lee proved in [34] that the Anosov theorem holds for homotopic periodic maps of infra-nilmanifolds and C. McCord extended this to homotopic periodic maps of infra-solvmanifolds ([45]). Note that f is a homotopic periodic map if some power of f is homotopic to the identity and infra-solvmanifolds are manifolds which have a finite regular cover by a solvmanifold. A final example are the virtual unipotent maps introduced and studied in [42] by W. Malfait. In Chapter 3 we examine the Anosov relation for expanding maps and the new class of nowhere expanding maps which form an extension of the class of virtual unipotent maps. In the fourth chapter we examine the Anosov diffeomorphisms.

Chapter 3

Periodic sequences and infra-nilmanifolds with an odd order holonomy group

To prove the results in this chapter[2] we introduce the concept of a periodic sequence associated to a map. This sequence is used in an analogue way in three different situations. Firstly, we show that the Anosov theorem can be generalized to the much larger class of infra-nilmanifolds with odd order holonomy group. Secondly we look at two classes of maps: the well-known expanding maps and the by us introduced nowhere expanding maps. This last class is the complete opposite of the expanding maps, as will be clear from the definition. For these nowhere expanding maps we can easily show that the Anosov relation holds. Finally we establish a necessary and sufficient condition for expanding maps f of infra-nilmanifolds M to satisfy the Anosov relation. Namely, $N(f) = |L(f)|$ if and only if M is orientable.

This last result is very interesting since we can use it to decide about the (non)-validity of the Anosov theorem for non-orientable infra-nilmanifolds. However, not every (non-orientable) infra-nilmanifold admits an expanding map. In fact, in the last section we present an example of a non-orientable infra-nilmanifold, which does not admit expanding maps and for which the Anosov theorem holds for this specific manifold.

3.1 The Anosov theorem for infra-nilmanifolds with odd order holonomy group

In this section we show that the Anosov theorem holds for all infra-nilmanifolds M with odd order holonomy group F. To be able to prove

---

[2] The results of this chapter can also be found in [15].

this statement, we will associate to any element of F a certain sequence of elements of F. The construction of this sequence depends on the map f for which we want to prove that $N(f) = |L(f)|$.

Suppose that $f : M \to M$ is a continuous map, that $(\delta, \mathfrak{D})$ is a homotopy lift of f and that $T : F \to \mathrm{Aut}(G)$ is the associated holonomy representation. Now we exploit the fact that $(\delta, \mathfrak{D})$ is obtained from Theorem 1.10, so we can apply Corollary 1.13.

Using this we can construct for any $x_1 \in F$ a sequence of elements of F, say $x_1, x_2, \ldots$, such that for any $i \in \mathbb{N}_0$: $\mathfrak{D}_* T_*(x_i) = T_*(x_{i+1}) \mathfrak{D}_*$. Since F is finite we know that we can construct this sequence in such a way that from a certain point onwards the sequence of elements becomes periodic. I.e. since F is finite we know that there has to exist a $j \geq 1$ and a $k \geq 1$ such that $x_{j+k} = x_j$. From that point we can take $x_{j+k+1} = x_{j+1}$, $x_{j+k+2} = x_{j+2}$, $\ldots$ (in general $x_{j+nk+l} = x_{j+l}$, for all $n, l \geq 0$) and still have the property that $\mathfrak{D}_* T_*(x_i) = T_*(x_{i+1}) \mathfrak{D}_*$, $\forall i \geq 1$. So, the sequence we construct for a given x_1 is periodic, with period k, from a certain position j onwards.

Note that we do not claim that this sequence is unique, but it is always possible to construct such a sequence as described above. Let us illustrate the set-up of such a periodic sequence in an example.

Example 3.1. *Let E be the Bieberbach group generated by*

$$(e_1, I_4), \cdots (e_4, I_4), (a, A) = (\begin{pmatrix} 1/4 \\ 0 \\ 0 \\ 0 \end{pmatrix}, \begin{pmatrix} 1 & 0 & 0 & 0 \\ 0 & 1 & 0 & 0 \\ 0 & 0 & 0 & -1 \\ 0 & 0 & 1 & 0 \end{pmatrix})$$

where e_i stands for a column vector of length 4 with 1 in the i-th place and 0 everywhere else ($1 \leq i \leq 4$). It can be verified that E is indeed a Bieberbach group which fits into the following short exact sequence

$$1 \to \mathbb{Z}^4 \to E \to \mathbb{Z}_4 \to 1.$$

So we obtain a 4-dimensional flat manifold $M = E \backslash \mathbb{R}^4$ with holonomy group $\mathbb{Z}_4$. Then take

$$(\delta, \mathfrak{D}) = (\begin{pmatrix} 0 \\ 0 \\ 0 \\ 0 \end{pmatrix}, \begin{pmatrix} 1 & \frac{1}{4} & 0 & 0 \\ -4 & 0 & 0 & 0 \\ 0 & 0 & 0 & 0 \\ 0 & 0 & 0 & 0 \end{pmatrix})$$

which induces a map $f : M \to M$ (this is verified as before). For this $(\delta, \mathfrak{D})$ we have that

$$(e_1 - 4e_2, I_4)(\delta, \mathfrak{D}) = (\delta, \mathfrak{D})(e_1, I_4);$$
$$(a, A)(\delta, \mathfrak{D}) = (\delta, \mathfrak{D})(e_2, I_4);$$
$$(-e_2 + a, A)(\delta, \mathfrak{D}) = (\delta, \mathfrak{D})(a, A)$$

Suppose that x generates the holonomy group of M, then we have that the periodic sequence associated to x is $x, x, x, \ldots$. On the other hand the periodic sequence for e is not unique: it can be taken equal to $e, e, e, \ldots$ but since $(a, A)(\delta, \mathfrak{D}) = (\delta, \mathfrak{D})(e_2, I_4)$ also equal to $e, x, x, \ldots$. As will be clear of the use of this periodic sequence, this non-uniqueness will not be a problem.

We will refer to the sequence $x_1, \ldots, x_j, \ldots, x_{j+k-1}, x_{j+k} = x_j, \ldots$ constructed as explained above as **a periodic sequence for x_1, associated to f, with period k starting from position j**. Using these sequences, we can now prove the following lemma.

Lemma 3.2. *Suppose M is an infra-nilmanifold with holonomy group F and associated holonomy representation $T : F \to \mathrm{Aut}(G)$. Suppose $f : M \to M$ is a continuous map and $(\delta, \mathfrak{D})$ is a homotopy lift of f. Let $x_1 \in F$ and let $x_1, x_2, x_3, \ldots$ be a periodic sequence for x_1 associated to f, with period k starting from position j. Then*

1. $\forall i \in \mathbb{N}_0$: $\det(I_n - T_*(x_1)\mathfrak{D}_*) = \det(I_n - T_*(x_i)\mathfrak{D}_*)$;
2. $\mathfrak{D}_*^k T_*(x_j) = T_*(x_j)\mathfrak{D}_*^k$;
3. $\exists l \in \mathbb{N}_0$: $(T_*(x_j)\mathfrak{D}_*)^l = \mathfrak{D}_*^l$.

<u>Proof:</u> We first prove that for any $i \geq 1$ we have that $\det(I_n - T_*(x_i)\mathfrak{D}_*) = \det(I_n - T_*(x_{i+1})\mathfrak{D}_*)$. Indeed,

$$\det(I_n - T_*(x_i)\mathfrak{D}_*) = \det(T_*(x_i))\det((T_*(x_i))^{-1} - \mathfrak{D}_*)$$
$$= \det(I_n - \mathfrak{D}_* T_*(x_i))$$
$$= \det(I_n - T_*(x_{i+1})\mathfrak{D}_*)$$

So for any element x_i of this periodic sequence we have that $\det(I_n - T_*(x_1)\mathfrak{D}_*) = \det(I_n - T_*(x_i)\mathfrak{D}_*)$.

To prove the second statement of this lemma, we compute

$$\mathfrak{D}_*^k T_*(x_j) = \mathfrak{D}_*^{k-1} T_*(x_{j+1}) \mathfrak{D}_*$$
$$= \mathfrak{D}_*^{k-2} T_*(x_{j+2}) \mathfrak{D}_*^2$$
$$= \ldots$$
$$= \mathfrak{D}_* T_*(x_{j+k-1}) \mathfrak{D}_*^{k-1}$$
$$= T_*(x_{j+k}) \mathfrak{D}_*^k$$
$$= T_*(x_j) \mathfrak{D}_*^k.$$

To obtain the third claim, let us first look at the k-th power of $T_*(x_j) \mathfrak{D}_*$.

$$(T_*(x_j) \mathfrak{D}_*)^k = T_*(x_j) \mathfrak{D}_* T_*(x_j) \mathfrak{D}_* (T_*(x_j) \mathfrak{D}_*)^{k-2}$$
$$= T_*(x_j) T_*(x_{j+1}) \mathfrak{D}_*^2 T_*(x_j) \mathfrak{D}_* (T_*(x_j) \mathfrak{D}_*)^{k-3}$$
$$= T_*(x_j) T_*(x_{j+1}) T_*(x_{j+2}) \mathfrak{D}_*^3 (T_*(x_j) \mathfrak{D}_*)^{k-3}$$
$$= \ldots$$
$$= T_*(x_j) T_*(x_{j+1}) \cdots T_*(x_{j+k-1}) \mathfrak{D}_*^k$$

Let $y = x_j x_{j+1} \cdots x_{j+k-1} \in F$. Then y is of finite order, say p, and analogously as before one can verify that $T_*(y)$ commutes with $\mathfrak{D}_*^k$. Therefore if $l = pk$, we obtain that $(T_*(x_j) \mathfrak{D}_*)^l = \mathfrak{D}_*^l$. □

With this lemma we can then prove the following theorem.

Theorem 3.3. *If M is an infra-nilmanifold with odd order holonomy group F, then $N(f) = |L(f)|$ for any continuous map $f : M \to M$.*

Proof: Suppose $f : M \to M$ is a continuous map of M and $(\delta, \mathfrak{D})$ is a homotopy lift of f. To use Theorem 2.5 we have to determine the sign of $\det(I_n - T_*(x_1) \mathfrak{D}_*)$ for any $x_1 \in F$. Construct a periodic sequence $x_1, \ldots, x_{j+k-1}, x_{j+k} = x_j, \ldots$ with period k starting from position j as explained above. Because of Lemma 3.2 it suffices to determine the sign of $\det(I_n - T_*(x_j) \mathfrak{D}_*^k)$ and we also know that $T_*(x_j) \mathfrak{D}_*^k = \mathfrak{D}_*^k T_*(x_j)$. We can pick a $P \in \mathrm{Gl}(n, \mathbb{R})$ which separates the eigenvalues of $\mathfrak{D}_*$:

$$P \mathfrak{D}_* P^{-1} = \begin{pmatrix} D_1 & 0 \\ 0 & D_2 \end{pmatrix}$$

with $D_1 \in M_m(\mathbb{R})$ (resp. $D_2 \in M_{n-m}(\mathbb{R})$) and for any eigenvalue λ of D_1 (resp. any eigenvalue λ of D_2) we have that $|\lambda| \leq 1$ (resp. $|\lambda| > 1$). The construction of D_1 and D_2 implies that D_1^k and D_2^k must have distinct eigenvalues and therefore we have that $T_*(x_j) \mathfrak{D}_*^k = \mathfrak{D}_*^k T_*(x_j)$ implies

$$PT_*(x_j)P^{-1} = \begin{pmatrix} T_1 & 0 \\ 0 & T_2 \end{pmatrix}$$

with $T_1 \in M_m(\mathbb{R})$ and $T_2 \in M_{n-m}(\mathbb{R})$. So for any $x_1 \in F$:

$$\det(I_n - T_*(x_1)\mathfrak{D}_*) = \det(I_n - T_*(x_j)\mathfrak{D}_*)$$
$$= \det(I_m - T_1 D_1)\det(I_{n-m} - T_2 D_2) \quad (3.1)$$

Lemma 3.2 guarantees the existence of a $l > 0$ such that $(T_1 D_1)^l = D_1^l$. This implies that for any eigenvalue λ of $T_1 D_1$, we also have that $|\lambda| \leq 1$. Let $\lambda_1, \lambda_2, \ldots, \lambda_m$ be the set of eigenvalues of $T_1 D_1$ (with each eigenvalue listed as many times as its algebraic multiplicity), then $|\lambda_i| \leq 1$ $(\forall i)$ implies that

$$\det(I_m - T_1 D_1) = (1 - \lambda_1) \cdots (1 - \lambda_m) \geq 0$$

So the first factor in (3.1) does not play a role in determining the sign of $\det(I_n - T_*(x_1)\mathfrak{D}_*)$.

Note that for any non-real eigenvalue λ_i, we know that $\overline{\lambda_i}$ is also an eigenvalue which implies that $(1 - \lambda_i)(1 - \overline{\lambda_i}) \geq 0$.

One can easily verify that the sign of the second factor depends on the number of real, positive eigenvalues of $T_2 D_2$ (note that Lemma 3.2 now implies that any eigenvalue λ of $T_2 D_2$ is of modulus > 1). Let us therefore denote the number (counted with multiplicity) of real, positive eigenvalues of D_2 by p and those of $T_2 D_2$ by q. If we can now show that $q \equiv p \bmod 2$, then all determinants must have the same sign and this would finish the proof.

Since F is of odd order, we have that $T_*(x_j)$ is of odd, finite order. Hence, also T_2 is of odd order which implies that $\det(T_2) = 1$. Therefore we have that

$$\det(T_2 D_2) = \det(T_2)\det(D_2) = \det(D_2). \quad (3.2)$$

Now, $\det(D_2) \neq 0$ since for any eigenvalue λ of D_2 yields that $|\lambda| > 1$. So the equality above a fortiori implies that both determinants have the same sign. So, modulo 2, the matrices D_2 and $T_2 D_2$ must have the same number of real, negative eigenvalues. But, as the number of non-real eigenvalues is even for both matrices, this also implies that $q \equiv p \bmod 2$. $\qquad \square$

Remark 3.4. *The condition that F is of odd order is crucial since for instance Anosov constructed in [1] a counter example on the Klein bottle (which is a flat manifold having $\mathbb{Z}_2$ as holonomy group). See also Chapter 5 for a detailed study of infra-nilmanifolds with cyclic holonomy groups or Part 3 for more counterexamples.*

3.2 Classes of maps for which the Anosov theorem holds

In the proof of Theorem 3.3 we separated the eigenvalues of the differential $\mathfrak{D}_*$ of the linear part of the homotopy lift. Because of this separation, we were able to construct matrices T_1 and T_2 (using the notations from the proof of Theorem 3.3). Now, in general, it is not possible to use the same techniques for infra-nilmanifolds with a holonomy group of even order. Since in that case -1 can be an eigenvalue of T_2 and so $\det(T_2)$ could be equal to -1. Moreover, because of the separation of $\mathfrak{D}_*$, we do not longer have control on how many times -1 appears as an eigenvalue of T_2.

In this section we avoid the separation of $\mathfrak{D}_*$ by looking at special classes of maps. Namely we assume that either D_1 or D_2 does not appear.

3.2.1 The Anosov relation for expanding maps

The first class of maps that we want to study are the expanding maps. Let us therefore first recall the definition.

Definition 3.5. *A C^1-self-map $f : M \to M$ on a closed smooth manifold M is said to be an expanding map if there exist constants $C > 0$ and $\mu > 1$ such that*

$$\|Df^n(v)\| \geq C\mu^n\|v\|, \forall v \in TM$$

for some Riemannian metric $\| \ \|$ on M.

It follows from the work of M. Gromov ([25]) that any expanding map of a compact manifold is topologically conjugated to an expanding infra-nilmanifold endomorphism. This means that for the homotopy lift $(\delta, \mathfrak{D})$ of an expanding map on an infra-nilmanifold we have that the modulus of all eigenvalues of $\mathfrak{D}_*$ is larger than 1.

The following theorem is based on the orientability of the infra-nilmanifold M which is completely determined by Proposition 1.6. Using this proposition we can now prove the following result about expanding maps of infra-nilmanifolds.

Theorem 3.6. *Suppose $f : M \to M$ is an expanding map of an infra-nilmanifold M. Then $N(f) = |L(f)|$ if and only if M is orientable.*

<u>Proof:</u> Suppose $f : M \to M$ is an expanding map and suppose $(\delta, \mathfrak{D})$ is a homotopy lift of f. We will use Theorem 2.5 and therefore we have to determine the sign of $\det(I_n - T_*(x_1)\mathfrak{D}_*)$ for any $x_1 \in F$. Construct a periodic sequence $x_1, \ldots, x_{j+k-1}, x_{j+k} = x_j, \ldots$ as before. Because of Lemma 3.2 is suffices to look at $\det(I_n - T_*(x_j)\mathfrak{D}_*)$ and we also know that $T_*(x_j)\mathfrak{D}_*^k = \mathfrak{D}_*^k T_*(x_j)$.

Let us first assume that M is orientable and so for any $x \in F$ we have that $\det(T_*(x)) = 1$. Therefore

$$\det(T_*(x_j)\mathfrak{D}_*) = \det(T_*(x_j))\det(\mathfrak{D}_*) = \det(\mathfrak{D}_*).$$

Recall that $\det(\mathfrak{D}_*) \neq 0$ since $|\lambda| > 1$ for any eigenvalue λ of $\mathfrak{D}_*$. Completely analogous as in the proof of Theorem 3.3 one can now verify that this implies that all the determinants have the same sign.

Let us now assume that M is non-orientable and so there exists an $x_1 \in F$ such that $\det(T_*(x_1)) = -1$. Then for any x_i of a periodic sequence $x_1, \ldots, x_{j+k-1}, x_{j+k} = x_j, \ldots$ we have that $\det(T_*(x_i)) = -1$. Indeed, by construction we have for any i that

$$\det(\mathfrak{D}_*)\det(T_*(x_i)) = \det(\mathfrak{D}_* T_*(x_i))$$
$$= \det(T_*(x_{i+1})\mathfrak{D}_*)$$
$$= \det(T_*(x_{i+1}))\det(\mathfrak{D}_*).$$

Since $\det(\mathfrak{D}_*) \neq 0$, this implies that $\det(T_*(x_i)) = \det(T_*(x_{i+1}))$ for any i.
So we can assume that $\det(T_*(x_j)) = -1$ and

$$\det(T_*(x_j)\mathfrak{D}_*) = \det(T_*(x_j))\det(\mathfrak{D}_*) = -\det(\mathfrak{D}_*).$$

Denote the number of positive, real eigenvalues of $\mathfrak{D}_*$ by p and of $T_*(x_j)\mathfrak{D}_*$ by q. Then one can, again analogously as in the proof of Theorem 3.3, easily verify that $q \not\equiv p \bmod 2$. This implies that $\det(I_n - \mathfrak{D}_*)$

and $\det(I_n - T_*(x_j)\mathfrak{D}_*)$ have a different sign. Note that these determinants can not be equal to zero, since for any eigenvalue λ of $\mathfrak{D}_* : |\lambda| > 1$ and because of the last statement of Lemma 3.2 the same holds for $T_*(x_j)\mathfrak{D}_*$. Theorem 2.5 then finishes the proof. $\qquad\square$

So this theorem implies that the Anosov relation for expanding maps is completely determined by the orientability of the infra-nilmanifold. It also implies that if M is a non-orientable manifold which admits an expanding map f, then the Anosov theorem no longer holds for M.

This is an interesting result since in [22] it was shown that all flat manifolds admit an expanding map. Moreover the same is true for 2-step infra-nilmanifolds ([36]), but on the other hand there are c-step infra-nilmanifolds (for $c > 2$) which do not admit expanding maps. See for instance [20] for a 3-step infra-nilmanifold which does not admit expanding maps. Since the above result will be a very useful result later on, we state it formally.

Proposition 3.7. *If M is a non-orientable flat manifold or a non-orientable 2-step infra-nilmanifold then the Anosov theorem does not hold for M.*

In general however this is no longer true and to prove this we construct a non-orientable infra-nilmanifold for which the Anosov theorem holds (and so a fortiori this manifold does not admit expanding maps). This example will be also useful in the following chapters and will show on more then one occasion that infra-nilmanifolds are much more complicated than flat manifolds. The example is presented in the last section of this chapter.

3.2.2 The Anosov relation for nowhere expanding maps

In this section we investigate the complete opposite of the class of expanding maps and introduce what we call the nowhere expanding maps.

Definition 3.8. *Let $f : M \to M$ be a continuous map of an infra-nilmanifold M, with a homotopy lift $(\delta, \mathfrak{D})$. Then f is said to be a nowhere expanding map, if $|\lambda| \le 1$ for all eigenvalues λ of $\mathfrak{D}_*$.*

Remark 3.9. *The class of nowhere expanding maps also contains the virtually unipotent maps (for which it is requested that all eigenvalues λ of $\mathfrak{D}_*$ are of modulus 1) introduced in [42].*

The class of nowhere expanding maps is in fact very closely related to the class of virtually unipotent maps. Indeed, one might be tempted to think that there are many possibilities for the moduli of the eigenvalues of $\mathfrak{D}_*$ since we only request them to be less than or equal to 1. However, the following result shows that in the case of nowhere expanding maps these moduli are either 1 or 0 and nothing in between is possible.

Lemma 3.10. *Let* $f : M \to M$ *be a nowhere expanding map on an infra-nilmanifold* M, *with homotopy lift* $(\delta, \mathfrak{D})$. *Let* λ *be an eigenvalue of* $\mathfrak{D}_*$, *then either* $|\lambda| = 1$ *or* $\lambda = 0$.

Proof: Let $f_* : E \to E$ be the homomorphism induced by f on the fundamental group E of $M = E\backslash G$. The group E is an almost-Bieberbach group, and we consider it as a subgroup of $G \rtimes \mathrm{Aut}(G)$. Now, let $\Lambda \subseteq G \cap E$ be a subgroup of finite index in E, satisfying $f_*(\Lambda) \subseteq \Lambda$. It follows from Theorem 1.10, that

$$\forall \lambda \in \Lambda : \ f_*(\lambda) = \delta \mathfrak{D}(\lambda) \delta^{-1} = (\mu(\delta) \circ \mathfrak{D})(\lambda)$$

where $\mu(\delta)$ denotes conjugation with δ.

We will now choose a special basis of the Lie algebra $\mathfrak{g}$ of G, by determining a specific generating set for the lattice Λ of G. Let $\gamma_1(\Lambda) = \Lambda$ and $\gamma_{i+1}(\Lambda) = [\Lambda, \gamma_i(\Lambda)]$. As Λ is a nilpotent group, $\gamma_{c+1}(\Lambda) = 1$ for some c. For any $i \geq 1$, we take

$$\Lambda_i = \sqrt{\gamma_i(\Lambda)} = \{x \in \Lambda \mid x^k \in \gamma_i(\Lambda), \text{ for some } k > 0\}.$$

Then Λ_i is a fully invariant subgroup of Λ, containing $\gamma_i(\Lambda)$ as a subgroup of finite index and we have that Λ_i/Λ_{i+1} is free abelian of finite rank, say k_i.

We now choose a set of generators

$$a_{1,1}, a_{1,2}, \ldots, a_{1,k_1}, a_{2,1}, \ldots, a_{2,k_2}, a_{3,1}, \ldots, a_{c,k_c}$$

for Λ, in such a way that $a_{i,j} \in \Lambda_i$, and the natural projections of $a_{i,1}, a_{i,2}, \ldots, a_{i,k_i}$ freely generate $\Lambda_i/\Lambda_{i+1} \cong \mathbb{Z}^{k_i}$.

As G is a nilpotent Lie group, the exponential map $\exp : \mathfrak{g} \to G$ is a diffeomorphism and we denote its inverse by log. The set $A_{1,1} = \log(a_{1,1}), \ldots, A_{c,k_c} = \log(a_{c,k_c})$ forms a basis of $\mathfrak{g}$. With respect to this basis, $(\mu(\delta) \circ \mathfrak{D})_*$ has a matrix representation of the form

$$\begin{pmatrix} B_1 & 0 & \cdots & 0 \\ * & B_2 & \cdots & 0 \\ \vdots & \vdots & \ddots & \vdots \\ * & * & \cdots & B_c \end{pmatrix}, \tag{3.3}$$

where B_i is a $(k_i \times k_i)$-matrix. In fact, B_i represents the morphism, induced by f_* on the free abelian group Λ_i/Λ_{i+1} and hence B_i is a matrix with integral entries. Now, as $\mathfrak{D}$ only differs from $\mu(\delta) \circ \mathfrak{D}$ by an inner conjugation of G, the matrix representing $\mathfrak{D}_*$ is also of the form (3.3) with the same integral blocks B_i on the diagonal (but with different entries below the diagonal). It follows that the eigenvalues of $\mathfrak{D}_*$ are exactly the same as the eigenvalues of the integral matrix

$$\begin{pmatrix} B_1 & 0 & \cdots & 0 \\ 0 & B_2 & \cdots & 0 \\ \vdots & \vdots & \ddots & \vdots \\ 0 & 0 & \cdots & B_c \end{pmatrix}.$$

The characteristic polynomial $p(x)$ of this matrix (and hence also of the matrix representing $\mathfrak{D}_*$) is of the form

$$p(x) = x^k(a_l x^l + \cdots + a_1 x + a_0)$$

with $a_i \in \mathbb{Z}$, $a_0 \neq 0$ and k is the multiplicity of the eigenvalue 0. This means that the non-zero eigenvalues $\lambda_1, \lambda_2, \cdots, \lambda_l$ satisfy $\lambda_1 \lambda_2 \cdots \lambda_l = a_0$. As it is given that $|\lambda_i| \leq 1$ ($1 \leq i \leq l$) and $|a_0| \geq 1$ (a_0 is a non-zero integer), we must have that $|\lambda_i| = 1$ ($1 \leq i \leq l$) (and $a_0 = \pm 1$). $\qquad \Box$

We are now ready to prove that the Anosov relation holds for any nowhere expanding map. This results generalizes Theorem 4.3 of [42], where the analogous result for virtually unipotent maps was obtained.

Theorem 3.11. *Let $f : M \to M$ be a nowhere expanding map on an infra-nilmanifold M, then $N(f) = L(f)$.*

Proof: Let $(\delta, \mathfrak{D})$ be a homotopy lift of f. We will use Theorem 2.5 and therefore we have to determine the sign of $\det(I_n - T_*(x_1)\mathfrak{D}_*)$ for any $x_1 \in F$. Construct a periodic sequence $x_1, \ldots, x_{j+k-1}, x_{j+k} = x_j, \ldots$ with period k and starting point j as explained in the previous section and suppose $\lambda_1, \ldots \lambda_n$ are the eigenvalues of $T_*(x_j)\mathfrak{D}_*$ (with again each eigenvalue listed as many times as its algebraic multiplicity). Because of Lemma 3.2 it suffices to look at $\det(I_n - T_*(x_j)\mathfrak{D}_*)$. Moreover the last statement of Lemma 3.2 implies that for any i, $1 \leq i \leq n$, we have that $|\lambda_i| \leq 1$ since for any eigenvalue λ of $\mathfrak{D}_* : |\lambda| \leq 1$. So for any $x_1 \in F$:

$$\det(I_n - T_*(x_1)\mathfrak{D}_*) = \det(I_n - T_*(x_j)\mathfrak{D}_*) = (1 - \lambda_1) \cdots (1 - \lambda_n) \geq 0.$$

Theorem 2.5 then finishes the proof. $\qquad \Box$

Example 3.12. *The map* f *introduced in Example 3.1 is an example of an nowhere expanding map since the eigenvalues of* $\mathfrak{D}$ *are* 0 *and* $\frac{1}{2} \pm i\frac{\sqrt{3}}{2}$ *and indeed* $\det(I_4 - A^k\mathfrak{D}) = 1 \geq 0$ *for any* k.

3.3 Infra-nilmanifolds are more complicated

In this section we construct a 7-dimensional non-orientable infra-nilmanifold M_1 with holonomy group $\mathbb{Z}_2$ for which the Anosov theorem holds (and hence M_1 does not admit any expanding map, see Proposition 3.7). This M_1 implies that the results obtained in the previous section are not true in general for k-step infra-nilmanifolds with $k > 2$. Let $\mathfrak{g}$ be the 7-dimensional Lie algebra with basis $X_1, X_2, \ldots, X_7$ where the non-zero Lie brackets between basis vectors are given by

$$[X_1, X_2] = X_3, \ [X_1, X_3] = X_4, \ [X_1, X_4] = X_5, \ [X_1, X_5] = X_6$$

$$[X_2, X_3] = -X_7, \ [X_2, X_7] = -X_5 - X_6, \ [X_3, X_7] = -X_6$$

Remark 3.13. *The reason for choosing this Lie algebra lies in the fact that we want to base our example on a Lie group with as few as possible automorphisms, certainly without any expanding automorphisms. The above Lie algebra is a characteristically nilpotent Lie algebra and hence any automorphism of this Lie algebra has only eigenvalues of modulus 1. Dimension 7 is the lowest possible dimension in which characteristically nilpotent Lie algebras occur. See [24, Chapter 2, section III] for more details.*

Note that we do not claim that there does not exist analogue examples in lower dimensions, since we did not examine this.

There is a faithful matrix representation of this Lie algebra, which is given by

$$\rho : \mathfrak{g} \to M_8(\mathbb{R}^n) : x_1 X_1 + x_2 X_2 + \cdots + x_7 X_7 \mapsto$$

$$\begin{pmatrix}
0 & x_1 & -\frac{2}{3}(x_2 + x_3) & 0 & \frac{x_7}{3} & \frac{x_7}{3} & 0 & x_6 \\
0 & 0 & -\frac{2}{3}x_2 & x_1 & 0 & \frac{x_7}{3} & 0 & x_5 \\
0 & 0 & 0 & 0 & -\frac{x_2}{2} & \frac{x_3}{2} & 0 & x_7 \\
0 & 0 & 0 & 0 & x_1 & 0 & 0 & x_4 \\
0 & 0 & 0 & 0 & 0 & x_1 & 0 & x_3 \\
0 & 0 & 0 & 0 & 0 & 0 & 0 & x_2 \\
0 & 0 & 0 & 0 & 0 & 0 & 0 & x_1 \\
0 & 0 & 0 & 0 & 0 & 0 & 0 & 0
\end{pmatrix}$$

One might use this matrix representation in checking the claims which follow.

Let $\exp : \mathfrak{g} \to G$ denote the exponential map from the nilpotent Lie algebra $\mathfrak{g}$ to the corresponding simply connected, connected nilpotent Lie group G. (Note that ρ lifts to a matrix representation of G). Consider

$$a_1 = \exp(X_1), \ a_2 = \exp(X_2), \ a_3 = \exp(\frac{1}{2}X_3), \ a_4 = \exp(\frac{1}{8}X_4),$$

$$a_5 = \exp(\frac{1}{48}X5), \ a_6 = \exp(\frac{1}{384}X_6) \text{ and } a_7 = \exp(\frac{1}{4}X_7).$$

Let $\mathfrak{T}$ be the automorphism of G, whose differential $\mathfrak{T}_* : \mathfrak{g} \to \mathfrak{g}$ satisfies:

$$\mathfrak{T}_*(X_1) = X_1, \ \mathfrak{T}_*(X_2) = -X_2, \ \mathfrak{T}_*(X_3) = -X_3, \ \mathfrak{T}_*(X_4) = -X_4,$$

$$\mathfrak{T}_*(X_5) = -X_5, \ \mathfrak{T}_*(X_6) = -X_6 \text{ and } \mathfrak{T}_*(X_7) = X_7.$$

Let $\alpha \in \mathrm{Aff}(G)$ be the element $(\exp(\frac{1}{2}X_1), \mathfrak{T})$. The subgroup E of $\mathrm{Aff}(G)$ generated by $a_1, \ldots, a_7$ and α has a presentation of the form

$$E = \langle a_1, a_2, \ldots, a_7, \alpha \mid [a_1, a_2] = a_3^2 a_4^{-4} a_5^{16} a_6^{112} a_7^2 \ [a_3, a_4] = 1 \qquad \rangle.$$

$$\begin{aligned}
&[a_1, a_3] = a_4^4 a_5^{-12} a_6^{32} && [a_3, a_5] = 1 \\
&[a_1, a_4] = a_5^6 a_6^{-24} && [a_3, a_6] = 1 \\
&[a_1, a_5] = a_6^8 && [a_3, a_7] = a_6^{-48} \\
&[a_1, a_6] = 1 && [a_4, a_5] = 1 \\
&[a_1, a_7] = 1 && [a_4, a_6] = 1 \\
&[a_2, a_3] = a_5^{-12} a_6^{-144} a_7^{-2} && [a_4, a_7] = 1 \\
&[a_2, a_4] = 1 && [a_5, a_6] = 1 \\
&[a_2, a_5] = 1 && [a_5, a_7] = 1 \\
&[a_2, a_6] = 1 && [a_6, a_7] = 1 \\
&[a_2, a_7] = a_5^{-12} a_6^{-96} \\
&\alpha^2 = a_1 && \alpha a_1 = a_1 \alpha \\
&\alpha a_2 = a_2^{-1} \alpha a_3 a_4^{-1} a_5^5 a_6^{47} a_7 \ \ \alpha a_3 = a_3^{-1} \alpha a_4^2 a_5^{-3} a_6^4 \\
&\alpha a_4 = a_4^{-1} \alpha a_5^3 a_6^{-6} && \alpha a_5 = a_5^{-1} \alpha a_6^4 \\
&\alpha a_6 = a_6^{-1} \alpha && \alpha a_7 = a_7 \alpha
\end{aligned}$$

This group E is an almost-crystallographic subgroup of $\mathrm{Aff}(G)$. Moreover, E is torsion-free and hence it is the fundamental group of an infra-nilmanifold M_1. The holonomy group is $\mathbb{Z}_2$ and the infra-nilmanifold is non-orientable ($\det(\mathfrak{T}_*) = -1$, see Proposition 1.6).

Consider now any self-map f of M_1, inducing an endomorphism θ on the fundamental group $\pi_1(M_1) = E$ and with homotopy lift $(\delta, \mathfrak{D})$. Then, we can distinguish two possibilities for $\mathfrak{D}_*$, namely $\mathfrak{D}_*\mathfrak{T}_* = \mathfrak{T}_*\mathfrak{D}_*$ or $\mathfrak{D}_*\mathfrak{T}_* = \mathfrak{D}_*$.

Any endomorphism $\mathfrak{D}_*$ of $\mathfrak{g}$ is completely determined by the images of its generators X_1 and X_2. Let us use the following notation

$$\mathfrak{D}_*(X_1) = \alpha_1 X_1 + \alpha_2 X_2 + \cdots + \alpha_7 X_7$$

and

$$\mathfrak{D}_*(X_2) = \beta_1 X_1 + \beta_2 X_2 + \cdots + \beta_7 X_7$$

Case 1: $\mathfrak{D}_*\mathfrak{T}_* = \mathfrak{T}_*\mathfrak{D}_*$ This case forces all of the parameters $\alpha_2, \alpha_3, \alpha_4,$ $\alpha_5, \alpha_6, \beta_1, \beta_7$ to be zero. Moreover, this case happens exactly when $\theta(\alpha) = a_1^{k_1} a_2^{k_2} \ldots a_7^{k_7} \alpha$ for some integers $k_i \in \mathbb{Z}$. It then follows that $\theta(a_1) = \theta(\alpha)^2 = a_1^{2k_1+1} a_2^{l_2} \cdots a_7^{l_7}$ for some other integers $l_i \in \mathbb{Z}$. This implies that $\alpha_1 = 2k_1 + 1 \neq 0$. It follows that in this case $\mathfrak{D}_*$ satisfies:

$$\mathfrak{D}_*(X_1) = (2k_1 + 1)X_1 + \alpha_7 X_7$$

and

$$\mathfrak{D}_*(X_2) = \beta_2 X_2 + \beta_3 X_3 + \cdots + \beta_6 X_6.$$

If we now require that $\mathfrak{D}_*$ actually determines an endomorphism of $\mathfrak{g}$, then there are 3 subcases to consider:

- $\beta_2 = \beta_3 = 0$ in which case $\mathfrak{D}_*$ has a matrix representation, w.r.t. the basis $X_1, X_2, \ldots, X_7$ of the form

$$\begin{pmatrix} 2k_1+1 & 0 & 0 & 0 & 0 & 0 & 0 \\ * & 0 & 0 & 0 & 0 & 0 & 0 \\ * & 0 & 0 & 0 & 0 & 0 & 0 \\ * & 0 & 0 & 0 & 0 & 0 & 0 \\ * & * & * & 0 & 0 & 0 & 0 \\ * & * & * & * & 0 & 0 & 0 \\ * & * & * & * & 0 & 0 & 0 \end{pmatrix}$$

It is obvious that $\det(I_7 - \mathfrak{D}_*) = \det(I_7 - \mathfrak{T}_*\mathfrak{D}_*) = -2k_1$ in this case.

- Another possibility is that $\beta_2 = \alpha_1 = 1$. Now $\mathfrak{D}_*$ is of the form

$$\begin{pmatrix} 1 & 0 & 0 & 0 & 0 & 0 & 0 \\ * & 1 & 0 & 0 & 0 & 0 & 0 \\ * & * & 1 & 0 & 0 & 0 & 0 \\ * & * & * & 1 & 0 & 0 & 0 \\ * & * & * & * & 1 & 0 & 0 \\ * & * & * & * & * & 1 & 0 \\ * & * & * & * & * & * & 1 \end{pmatrix}$$

which again implies that $\det(I_7 - \mathfrak{D}_*) = \det(I_7 - \mathfrak{T}_*\mathfrak{D}_*) = 0$.

- Finally it is possible that $\alpha_1 = 1$ and $\beta_2 = -1$. Now, $\mathfrak{D}_*$ is of the form

$$\begin{pmatrix} 1 & 0 & 0 & 0 & 0 & 0 & 0 \\ * & -1 & 0 & 0 & 0 & 0 & 0 \\ * & * & -1 & 0 & 0 & 0 & 0 \\ * & * & * & -1 & 0 & 0 & 0 \\ * & * & * & * & -1 & 0 & 0 \\ * & * & * & * & * & -1 & 0 \\ * & * & * & * & * & * & 1 \end{pmatrix}$$

which again implies that $\det(I_7 - \mathfrak{D}_*) = \det(I_7 - \mathfrak{T}_*\mathfrak{D}_*) = 0$.

Case 2: $\mathfrak{D}_*\mathfrak{T}_* = \mathfrak{D}_*$ This case forces $\beta_1 = \beta_2 = \cdots = \beta_7 = 0$. Hence $\mathfrak{D}_*$ is of the form

$$\begin{pmatrix} k_1 & 0 & 0 & 0 & 0 & 0 & 0 \\ k_2 & 0 & 0 & 0 & 0 & 0 & 0 \\ k_3 & 0 & 0 & 0 & 0 & 0 & 0 \\ k_4 & 0 & 0 & 0 & 0 & 0 & 0 \\ k_5 & 0 & 0 & 0 & 0 & 0 & 0 \\ k_6 & 0 & 0 & 0 & 0 & 0 & 0 \\ k_7 & 0 & 0 & 0 & 0 & 0 & 0 \end{pmatrix}$$

Again we obtain that $\det(I_7 - \mathfrak{D}_*) = \det(I_7 - \mathfrak{T}_*\mathfrak{D}_*) = 1 - k_1$.

As before we can use Theorem 2.5 to conclude that the Anosov theorem holds on this infra-nilmanifold.

Note that this already shows in two ways that infra-nilmanifolds are much more complicated then flat manifolds. Firstly, there are no non-orientable flat manifolds for which the Anosov theorem holds.

Secondly, in Chapter 5 we will show that the Anosov theorem never holds for flat manifolds with $\mathbb{Z}_2$ as holonomy group. This example shows that this no longer holds in general for infra-nilmanifolds with $\mathbb{Z}_2$ as holonomy group.

Chapter 4

Anosov diffeomorphisms

In the previous chapter we examined two ('opposite') classes of maps which could be defined based on their eigenvalues of the homotopy lifts. In this chapter we examine another class of maps, namely the well studied class of Anosov diffeomorphisms. This class differs from what was studied in the previous chapter because, as we show later on, for these maps we always have both a D_1 and a D_2 (using the notations of the proof of Theorem 3.3). To prove this we start this chapter by recalling an algebraic characterization of Anosov diffeomorphisms.

A motivation for the examination of the Anosov relation for Anosov diffeomorphisms is the fact that these maps play an important role in dynamics. They could for instance be used to form nice examples of structurally stable dynamical systems, which also implies that $MF(f)$ is very interesting information. However it is important to note that not every compact manifold admits Anosov diffeomorphisms. Up till now the only known examples of Anosov diffeomorphisms are maps of infra-nilmanifolds. Moreover it has been conjectured that every Anosov diffeomorphism is topologically conjugated to an Anosov diffeomorphism of an infra-nilmanifold (see [43]) .

H. Porteous gave in [49] a complete characterization of the flat manifolds admitting Anosov diffeomorphisms. Unfortunately there is no such characterization for infra-nilmanifolds in general. Moreover J. Lauret argued in [35] that a reasonable classification of nilmanifolds admitting Anosov diffeomorphisms, and so a fortiori also of infra-nilmanifolds, is not possible. Therefore we focus on Anosov diffeomorphisms of n-dimensional flat manifolds.

In [41], it was proved that if M admits Anosov diffeomorphisms, then its first Betti number $b_1(M)$ satisfies one of the following: $b_1(M) = 0$, $2 \leq b_1(M) \leq n-2$ or $b_1(M) = n$ (and all situations occur). In the last case M is a torus and so $N(f) = |L(f)|$ for every continuous self-map f of M. For the other cases, we investigate the possibility of constructing a flat manifold M, with prescribed first Betti number, such that on the one hand M admits an Anosov diffeomorphism f with $N(f) \neq |L(f)|$ and on the other hand M also supports an Anosov diffeomorphism g satisfying $N(g) = |L(g)|$.

In the second section we show that for the case of non-primitive flat manifolds M, i.e. $b_1(M) \neq 0$, it turns out that this is always possible except when $b_1(M) = n-2$ (Theorem 4.6). In this latter case we show that $N(f) \neq |L(f)|$ for each flat manifold M (admitting an Anosov diffeomorphism) and each Anosov diffeomorphism f on M (Theorem 4.10). Primitive flat manifolds which admit Anosov diffeomorphisms only exist from dimension 6 onwards ([49]). In the third section we will construct such primitive flat manifolds and the desired Anosov diffeomorphisms f and g in any dimension $n > 6$ (Theorem 4.14). For each 6-dimensional primitive flat manifold M admitting an Anosov diffeomorphism however, we show that $N(f) = |L(f)|$ for all Anosov diffeomorphisms f of M (Theorem 4.16).

Finally we stress that we do not present an exhaustive examination of all possible Anosov diffeomorphisms of flat manifolds. Based on the obtained results, we obviously expect for $f : M \to M$ of flat manifolds that the property of f being an Anosov diffeomorphism has not much influence on f satisfying the Anosov relation. Except for some boundary cases for which we then present an exhaustive examination.[3]

4.1 Algebraic characterization

Let $f : M \to M$ be a diffeomorphism of a flat n-dimensional manifold M. We say that f is an affine diffeomorphism of M if the lifting of f to the universal cover $\mathbb{R}^n$ of M belongs to $\mathrm{Aff}(\mathbb{R}^n)$. In fact, such a lifting then automatically belongs to the normalizer $N_{\mathrm{Aff}(\mathbb{R}^n)}(E)$ of E. An affine diffeomorphism f of M is hyperbolic if and only if f lifts to an affine transformation $(a, A) \in \mathrm{Aff}(\mathbb{R}^n)$ with A hyperbolic (i.e. having no eigenvalues of absolute value 1).

---

[3] The results of this chapter can also be found in [14].

Definition 4.1. *A diffeomorphism* $f : M \to M$ *of a closed smooth manifold is an Anosov diffeomorphism if and only if tangent space TM decomposes as a direct sum of a contracting and an expanding part. That is, there is a continuous splitting $TM = E^s \oplus E^u$ such that for some (and hence any) any Riemannian metric on M, there are constants c and λ ($c > 0, 1 < \lambda$) such that for all $r \in \mathbb{N}_0$, $\|df^r(v)\| \geq c\lambda^r\|v\|$ (for all $v \in E^u$) and $\|df^r(w)\| \leq c^{-1}\lambda^{-r}\|w\|$ (for all $w \in E^s$).*

Such maps are studied in e.g. [23] and [43]. The examination of Anosov diffeomorphisms of flat manifolds can be converted into a pure algebraic problem. Namely, each hyperbolic diffeomorphism of a flat Riemannian manifold M defines an Anosov diffeomorphisms on M and conversely, we show in the following lemma that each Anosov diffeomorphism of M is homotopic to a hyperbolic diffeomorphism of M.

Lemma 4.2. *If $f : M \to M$ is an Anosov diffeomorphism of a flat manifold M, then f is homotopic to a hyperbolic diffeomorphism $g : M \to M$.*

<u>Proof:</u> In [43] Manning showed that each Anosov diffeomorphism f of M is topologically conjugated with a hyperbolic diffeomorphism g of M. Thus there exists a homeomorphism $h : M \to M$ such that $f = hgh^{-1}$. Since h is a homeomorphism of M, the induced map $h_* : \pi_1(M) \to \pi_1(M)$ is an isomorphism. Then because of the second theorem of Bieberbach there exists a $(\delta, \mathfrak{D}) \in \mathrm{Aff}(\mathbb{R}^n)$ such that h_* is exactly the conjugation with $(\delta, \mathfrak{D})$ restricted to $\pi_1(M)$. This $(\delta, \mathfrak{D})$ induces an affine diffeomorphism $\tilde{h}$ on M. Since M is aspherical, h and $\tilde{h}$ are homotopic ([54, page 225]). So f is homotopic to $\tilde{h}g\tilde{h}^{-1}$. Suppose g is induced by a $(\delta_1, \mathfrak{D}_1) \in \mathrm{Aff}(\mathbb{R}^n)$ with $\mathfrak{D}_1$ hyperbolic, then $\mathfrak{D}\mathfrak{D}_1\mathfrak{D}^{-1}$ is also hyperbolic and $\tilde{h}g\tilde{h}^{-1}$ is a hyperbolic diffeomorphism. $\square$

A complete characterization of flat Riemannian manifolds supporting Anosov diffeomorphisms, due to H. Porteous ([49]), is the following:

Theorem 4.3. *A n-dimensional flat manifold M with holonomy group F and associated holonomy representation $T : F \to \mathrm{Gl}(n, \mathbb{Z})$ admits an Anosov diffeomorphism if and only if each $\mathbb{Q}$-irreducible component of T of multiplicity one is reducible over $\mathbb{R}$.*

Finally we present a criterion to calculate the first Betti number of a flat n-dimensional manifold M. Since we can see $\pi_1(M)$ as a subgroup of $\mathrm{Aff}(\mathbb{R}^n) = \mathbb{R}^n \rtimes \mathrm{Gl}(n, \mathbb{R})$, we may assume that $\pi_1(M)$ is generated by

$(a_1, A_1), \ldots, (a_k, A_k)$. The first Betti number $b_1(M)$ of M, or equivalently the first Betti number $b_1(E)$ of E, is by definition the rank of the first homology group $H_1(M, \mathbb{Z})$ (or equivalently of $H_1(E, \mathbb{Z})$). Because of [12, Remark 6.4.16] we know that

$$b_1(M) = n - rank(A_1 - I_n, \ldots, A_k - I_n),$$

4.2 Non-primitive flat manifolds

As mentioned already in the introduction, if a flat manifold M admits an Anosov diffeomorphism, its first Betti number $b_1(M)$ must satisfy $b_1(M) = 0$, $2 \leq b_1(M) \leq n - 2$ or $b_1(M) = n$. In this section we consider the case $2 \leq b_1(M) \leq n - 2$ and we do this by first looking at $b_1(M) < n - 2$ and then at $b_1(M) = n - 2$.

4.2.1 Flat n-dimensional manifolds with first Betti number smaller than $n - 2$

For any integer $n \geq 2$ and any integer k satisfying $1 \leq k \leq n - 1$, we define the group $E_{n,k}$ generated by

$$(e_i, I_n) \text{ and } (a, A) = (\begin{pmatrix} \frac{1}{2} \\ 0 \\ \vdots \\ 0 \end{pmatrix}, \begin{pmatrix} I_k & 0 \\ 0 & -I_{n-k} \end{pmatrix}))$$

with 1 in the i-th place and 0 everywhere else for e_i ($1 \leq i \leq n$). One easily verifies that $E_{n,k}$ is torsion-free and $\mathbb{Z}^n$ is maximal abelian in $E_{n,k}$, because the associated holonomy representation clearly is faithful. Thus $E_{n,k}$ is a Bieberbach group fitting into the following short exact sequence:

$$1 \to \mathbb{Z}^n \to E_{n,k} \to \mathbb{Z}_2 \to 1.$$

$M_{n,k} = E_{n,k} \backslash \mathbb{R}^n$ is a n-dimensional flat manifold with first Betti number equal to k. The affine diffeomorphisms of $M_{n,k}$ are exactly those diffeomorphisms of $M_{n,k}$ lifting to a $(\delta, \mathfrak{D}) \in \mathrm{Aff}(\mathbb{R}^n)$ and normalizing $E_{n,k}$. In the next lemma we therefore characterize $N_{\mathrm{Aff}(\mathbb{R}^n)}(E_{n,k})$.

Lemma 4.4. *Suppose* $(\delta, \mathfrak{D}) \in \mathrm{Aff}(\mathbb{R}^n)$, *then*

$$(\delta, \mathfrak{D}) \in N_{\mathrm{Aff}(\mathbb{R}^n)}(E_{n,k}) \Leftrightarrow \delta = \begin{pmatrix} x_1 \\ \vdots \\ x_n \end{pmatrix} \ and \ \mathfrak{D} = (d_{ij}) = \begin{pmatrix} D_1 & 0 \\ 0 & D_2 \end{pmatrix}$$

with $x_1, \ldots, x_k \in \mathbb{R}$, $x_{k+1}, \ldots x_n \in \frac{1}{2}\mathbb{Z}$, $D_1 \in \mathrm{Gl}(k, \mathbb{Z})$, $D_2 \in \mathrm{Gl}(n - k, \mathbb{Z})$ and $d_{11} \in 1 + 2\mathbb{Z}$, $d_{21}, \ldots d_{k1} \in 2\mathbb{Z}$.

<u>Proof:</u> Suppose $\delta = (x_1, \ldots, x_n)^t$ with $x_i \in \mathbb{R}$ and $\mathfrak{D} = (d_{ij}) \in \mathrm{Gl}(n, \mathbb{R})$. First we consider the conjugates of (e_i, I_n):

$$\begin{aligned} (\delta, \mathfrak{D})(e_i, I_n)(\delta, \mathfrak{D})^{-1} &= (\delta + \mathfrak{D}e_i, \mathfrak{D})(-\mathfrak{D}^{-1}\delta, \mathfrak{D}^{-1}) \\ &= (\mathfrak{D}e_i, I_n). \end{aligned}$$

This belongs to $E_{n,k}$ if and only if all the elements of $\mathfrak{D}e_i$ are integers. Since this can be repeated for each i, $1 \leq i \leq n$, and $(\delta, \mathfrak{D})^{-1}$ is also in $N_{\mathrm{Aff}(\mathbb{R}^n)}(E_{n,k})$ this means that $\mathfrak{D} \in \mathrm{Gl}(n, \mathbb{Z})$.
Analogously for (a, A):

$$\begin{aligned} (\delta, \mathfrak{D})(a, A)(\delta, \mathfrak{D})^{-1} &= (\delta + \mathfrak{D}a, \mathfrak{D}A)(-\mathfrak{D}^{-1}\delta, \mathfrak{D}^{-1}) \\ &= (\delta + \mathfrak{D}a - \mathfrak{D}A\mathfrak{D}^{-1}\delta, \mathfrak{D}A\mathfrak{D}^{-1}). \end{aligned}$$

To obtain an element of $E_{n,k}$, $\mathfrak{D}A\mathfrak{D}^{-1}$ should be equal to A. This implies that $\mathfrak{D}$ must be of the form $\begin{pmatrix} D_1 & 0 \\ 0 & D_2 \end{pmatrix}$ where $D_1 \in \mathrm{Gl}(k, \mathbb{Z})$ and $D_2 \in \mathrm{Gl}(n - k, \mathbb{Z})$. This is however not sufficient. The translational part

$$\delta + \mathfrak{D}a - A\delta = (\frac{1}{2}d_{11}, \ldots, \frac{1}{2}d_{k1}, 2x_{k+1}, \ldots, 2x_n)^t$$

must be equal to $(\frac{1}{2} + k_1, k_2, \ldots, k_n)^t$ for some $k_i \in \mathbb{Z}$. We conclude that $x_1, \ldots, x_k \in \mathbb{R}$, $x_{k+1}, \ldots x_n \in \frac{1}{2}\mathbb{Z}$, $d_{11} \in 1 + 2\mathbb{Z}$ and $d_{21}, \ldots d_{k1} \in 2\mathbb{Z}$. $\square$

In order to construct Anosov diffeomorphisms on $M_{n,k}$, it suffices to choose a pair $(\delta, \mathfrak{D})$ as in Lemma 4.4 with $\mathfrak{D}$ hyperbolic. Note that for each $n \geq 4$ and for any k with $2 \leq k \leq n - 2$ such a $\mathfrak{D}$ indeed exists. One can always construct a blocked diagonal matrix using the blocks $\begin{pmatrix} 1 & 1 \\ 2 & 1 \end{pmatrix}$ and $\begin{pmatrix} 1 & 1 & 1 \\ 2 & 1 & 1 \\ 2 & 2 & 3 \end{pmatrix}$.

In the sequel we need the following technical lemma.

Lemma 4.5.

1. *For every hyperbolic $H_2 \in \mathrm{Gl}(2, \mathbb{R})$ with $\det(H_2) = \pm 1$, we have*

$$\mathrm{sgn}(\det(I_2 - H_2)) \neq \mathrm{sgn}(\det(I_2 + H_2)).$$

2. *For every integer $n \geq 3$ there exist hyperbolic $H_n, H'_n \in \mathrm{Gl}(n, \mathbb{Z})$ such that*

$$\mathrm{sgn}(\det(I_n - H_n)) \neq \mathrm{sgn}(\det(I_n + H_n));$$
$$\mathrm{sgn}(\det(I_n - H'_n)) = \mathrm{sgn}(\det(I_n + H'_n)).$$

<u>Proof:</u>

1. Suppose $H_2 = (h_{ij}) \in \mathrm{Gl}(2, \mathbb{R})$ is hyperbolic and $\det(H_2) = \pm 1$. Then

$$\det(I_2 - H_2) = 1 - (h_{11} + h_{22}) + \det(H_2),$$
$$\det(I_2 + H_2) = 1 + (h_{11} + h_{22}) + \det(H_2)$$

and the eigenvalues of H_2 are given by

$$\frac{1}{2}\left(h_{11} + h_{22} \pm \sqrt{(h_{11} + h_{22})^2 - 4\det(H_2)}\right).$$

Since $\det(H_2) = \pm 1$, we can distinguish two cases:
   a) If $\det(H_2) = -1$, then $\det(I_2 - H_2) = -(h_{11} + h_{22})$ and $\det(I_2 + H_2) = h_{11} + h_{22}$. Thus their sign can only be equal if $h_{11} + h_{22} = 0$. But in that case the eigenvalues of H_2 are of absolute value 1, contradicting the hyperbolicity of H_2.
   b) If $\det(H_2) = 1$, then $\det(I_2 - H_2) = 2 - (h_{11} + h_{22})$ and $\det(I_2 + H_2) = 2 + (h_{11} + h_{22})$. Their sign can only be equal if

$$-2 \leq h_{11} + h_{22} \leq 2.$$

But then $(h_{11} + h_{22})^2 - 4\det(H_2) \leq 0$ and so for the eigenvalues λ_1 and λ_2 of H_2 we obtain $\lambda_1 = \overline{\lambda_2}$. Then again the eigenvalues of H_2 are of absolute value 1, since their product must be equal to 1.

2. In order to prove the second part, we first construct hyperbolic H_2, H_3 and H_4 such that $\mathrm{sgn}(\det(I_k - H_k)) \neq \mathrm{sgn}(\det(I_k + H_k))$ ($k = 2, 3, 4$) and hyperbolic H'_3 and H'_4 with $\mathrm{sgn}(\det(I_k - H'_k)) =$

$\text{sgn}(\det(I_k + H'_k))$ (for $k = 3$ or 4). For H_2 one can take any hyperbolic matrix in $\text{Gl}(2,\mathbb{Z})$ by the first step of this lemma, e.g. $H_2 = \begin{pmatrix} 1 & 1 \\ 2 & 1 \end{pmatrix}$, for the matrices H_3, H'_3, H_4 and H'_4 one can take:

$$H_3 = \begin{pmatrix} 1 & 1 & 1 \\ 2 & -1 & 3 \\ 2 & 0 & 3 \end{pmatrix} \quad H'_3 = \begin{pmatrix} 1 & 1 & 1 \\ 2 & 1 & 1 \\ 2 & 2 & 3 \end{pmatrix},$$

$$H_4 = \begin{pmatrix} 1 & 1 & 1 & 1 \\ 2 & 1 & 1 & 1 \\ 2 & 2 & 1 & 1 \\ 2 & 2 & 2 & 3 \end{pmatrix} \quad H'_4 = \begin{pmatrix} 1 & 1 & 0 & 0 \\ 2 & 1 & 0 & 0 \\ 0 & 0 & 1 & 1 \\ 0 & 0 & 2 & 1 \end{pmatrix}.$$

For every integer $n \geq 3$ we now construct H_n and H'_n as a blocked diagonal matrix in $\text{Gl}(n,\mathbb{Z})$ consisting of blocks inside $\text{Gl}(2,\mathbb{Z})$, $\text{Gl}(3,\mathbb{Z})$ and $\text{Gl}(4,\mathbb{Z})$. Indeed, in order to find a general H_n, we can then consider a blocked diagonal matrix consisting of one block H_k ($2 \leq k \leq 4$) together with the right amount of blocks H'_3. On the other hand, H'_5 can be build using one block H_2 and one block H_3, H'_6 using two blocks H'_3, H'_7 using one H'_3 and one H'_4, ... $\square$

Now we prove the following theorem

Theorem 4.6. *For each dimension $n > 4$ and for each integer k, satisfying $2 \leq k < n - 2$, there exists a flat manifold M with $b_1(M) = k$ and admitting Anosov diffeomorphisms f, g such that $N(f) \neq |L(f)|$ and $N(g) = |L(g)|$.*

<u>Proof:</u> We work with the flat manifold $M_{n,k}$ constructed in the beginning of this section and the $(\delta, \mathfrak{D}) = (\delta, \begin{pmatrix} D_1 & 0 \\ 0 & D_2 \end{pmatrix}) \in N_{\text{Aff}(\mathbb{R}^n)}(E_{n,k})$ obtained in Lemma 4.4. To prove our statement concerning the Nielsen and Lefschetz number of f and g we use Theorem 2.5. So we have to calculate $\det(I_n - \mathfrak{D})$ and $\det(I_n - A\mathfrak{D})$.

$$\det(I_n - \mathfrak{D}) = \det \begin{pmatrix} I_k - D_1 & 0 \\ 0 & I_{n-k} - D_2 \end{pmatrix}$$
$$= \det(I_k - D_1)\det(I_{n-k} - D_2)$$

$$\det(I_n - A\mathfrak{D}) = \det \begin{pmatrix} I_k - D_1 & 0 \\ 0 & I_{n-k} + D_2 \end{pmatrix}$$
$$= \det(I_k - D_1)\det(I_{n-k} + D_2)$$

Notice that $\det(I_n - \mathfrak{D}) = 0$ or $\det(I_n - A\mathfrak{D}) = 0$ only occurs if 1 or -1 is an eigenvalue, which is not possible for hyperbolic diffeomorphisms. Because of Lemma 4.5 and the fact that $n - k \geq 3$, we know that we can find suitable D_2 and D_2' such that in the first case $\mathrm{sgn}(\det(I_n - \mathfrak{D})) = -\mathrm{sgn}(\det(I_n - A\mathfrak{D}))$ and in the second case $\mathrm{sgn}(\det(I_n - \mathfrak{D}')) = \mathrm{sgn}(\det(I_n - A\mathfrak{D}'))$. So with δ, D_1 as in Lemma 4.4,

1. $D_2 = H_{n-k}$ as in Lemma 4.5 and $f : M_{n,k} \to M_{n,k}$ induced by $(\delta, \begin{pmatrix} D_1 & 0 \\ 0 & D_2 \end{pmatrix})$, we have that $|L(f)| \neq N(f)$;

2. $D_2' = H_{n-k}'$ as in Lemma 4.5 and $g : M_{n,k} \to M_{n,k}$ induced by $(\delta, \begin{pmatrix} D_1 & 0 \\ 0 & D_2' \end{pmatrix})$, we have that $|L(g)| = N(g)$.

$\square$

Remark 4.7. *The first part of this theorem is a generalization of a theorem due to S. Kwasik and K.B. Lee ([34]) proving the existence of such Anosov diffeomorphisms in each even dimension.*

4.2.2 Flat manifolds with first Betti number equal to $n - 2$

We first prove the following lemma concerning the first Betti number.

Lemma 4.8. *Suppose M is a flat n-dimensional manifold with first Betti number $b_1(M) = k > 0$. Assume that $E = \pi_1(M)$ is its fundamental group and*

$$T : F \to \mathrm{Gl}(n, \mathbb{Z})$$

is its holonomy representation. Then T is $\mathbb{Q}$-equivalent to a representation of the form

$$T' : F \to \mathrm{Gl}(n, \mathbb{Z}) : x \mapsto \begin{pmatrix} I_k & 0 \\ 0 & B(x) \end{pmatrix},$$

with $B(x) \in \mathrm{Gl}(n - k, \mathbb{Z})$, $\forall x \in F$.

Proof: To prove this, let F be generated by $x_1, x_2, \ldots x_l$ and let $T(x_i) = A_i$ ($1 \leq i \leq l$). For any i we define the set $V_i = \{\overrightarrow{v} \in \mathbb{Q}^n | \overrightarrow{v} A_i = \overrightarrow{v}\}$ and take $V = A_1 \cap A_2 \cap \ldots \cap A_l$. We know that $b_1(M) = k$ if and only if

$$k = n - rank(A_1 - I_n, \ldots, A_l - I_n)$$

or $rank(A_1 - I_n, \ldots, A_l - I_n) = n - k$. So V is a k-dimensional rational vector space.

Let us now define $Z_V = \mathbb{Z}^n \cap V$. Then Z_V is a free abelian subgroup of $\mathbb{Z}^n$ of rank k and $\mathbb{Z}^n / Z_V$ is torsion-free. Therefore we can determine a new set of free generators for $\mathbb{Z}^n$ by first choosing k generators for Z_V and then adding $n - k$ elements such that their canonical projections generate $\mathbb{Z}^n / Z_V$.

With respect to such a set of generators, determined by a change of basis matrix $P \in \text{Gl}(n, \mathbb{Z})$, we find that the holonomy representation is now given as

$$T_1 : F \to \text{Gl}(n, \mathbb{Z}) : x \mapsto PT(x)P^{-1} = \begin{pmatrix} I_k & 0 \\ C(x) & B(x) \end{pmatrix}$$

with $B(x) \in \text{Gl}(n - k, \mathbb{Z})$ and $C(x)$ a $(n - k) \times k$ matrix with integral entries. As F is a finite group and $\mathbb{Q}$ is a field of characteristic 0, we know that any representation of F over $\mathbb{Q}$ is fully reducible. In fact there exists a matrix $Q \in \text{Gl}(n, \mathbb{Q})$, such that $T' = QT_1Q^{-1}$ is of the form

$$T' : F \to \text{Gl}(n, \mathbb{Z}) : x \mapsto \begin{pmatrix} I_k & 0 \\ 0 & B(x) \end{pmatrix}.$$

This finishes the proof of the lemma. □

Let us now return to the case $b_1(M) = n - 2$. The flat manifold $M_{n,2}$ with fundamental group $E_{n,2}$ generated by the translations

$(e_1, I_n), \ldots, (e_n, I_n)$ and $(a_1, A_1) = (\begin{pmatrix} \frac{1}{2} \\ \vdots \\ 0 \\ 0 \end{pmatrix}, \begin{pmatrix} I_{n-2} & 0 \\ 0 & -I_2 \end{pmatrix}))$ is an exam-

ple of a manifold such that $b_1(M_{n,2}) = n - 2$. Moreover, because of Theorem 4.3, we know that this manifold admits Anosov diffeomorphisms. Concerning $M_{n,2}$ and $E_{n,2}$ we can prove the following proposition:

Proposition 4.9. Let $n \geq 4$. Suppose M is a n-dimensional flat manifold with $b_1(M) = n - 2$ and $\pi_1(M) = E$. Then M admits Anosov diffeomorphisms if and only if E is in the same $\mathbb{Q}$-class as $E_{n,2}$.

Being in the same $\mathbb{Q}$-class means that the holonomy representations of E and $E_{n,2}$ are $\mathbb{Q}$-equivalent.

Proof: It is an immediate consequence of Theorem 4.3 and the fact that $M_{n,2}$ admits Anosov diffeomorphisms, that any flat manifold M whose fundamental group is the same $\mathbb{Q}$-class as $E_{n,2}$ also admits Anosov diffeomorphisms.

Conversely assume that M admits Anosov diffeomorphisms. Because of Lemma 4.8, we know that the holonomy representation of E is $\mathbb{Q}$-equivalent to a representation of the form

$$T' : F \to \mathrm{Gl}(n,\mathbb{Z}) : x \mapsto \begin{pmatrix} I_{n-2} & 0 \\ 0 & B(x) \end{pmatrix}$$

with $B(x) \in \mathrm{Gl}(2,\mathbb{Z})$. Moreover, as the holonomy representation is faithful the representation $B : F \to \mathrm{Gl}(2,\mathbb{Z}) : x \mapsto B(x)$ has to be faithful too. There are only 10 $\mathbb{Q}$-classes of such representations (see [3]). From these 10 $\mathbb{Q}$-classes of representations, there are only two of them satisfying Theorem 4.3, the trivial one (for the trivial group) and the representation $\varphi : \mathbb{Z}_2 \to \mathrm{Gl}(2,\mathbb{Z})$, mapping the non-identity element of $\mathbb{Z}_2$ onto $-I_2$. We have to exclude the trivial group F, for in this case M is a torus and has $b_1(M) = n$. Therefore, the only possibility is that the holonomy group of M is $\mathbb{Z}_2$ and its holonomy representation is $\mathbb{Q}$-equivalent to that of $M_{n,2}$.   □

Now we can prove the theorem stated at the beginning of this section.

Theorem 4.10. *Suppose $n \geq 4$ and let M be a n-dimensional flat manifold with $b_1(M) = n - 2$. Then, for each Anosov diffeomorphism $f : M \to M$:*

$$N(f) \neq |L(f)|.$$

Proof: Let $E = \pi_1(M)$ be the fundamental group of M. By Proposition 4.9, we know that the holonomy group of M is $\mathbb{Z}_2$. So E is generated by n translations and an element (a, A), where $A \in \mathrm{Gl}(n,\mathbb{Z})$ is of order two. Again by Proposition 4.9, we know that there exists a matrix $P \in \mathrm{Gl}(n,\mathbb{Q})$ such that $PAP^{-1} = \begin{pmatrix} I_{n-2} & 0 \\ 0 & -I_2 \end{pmatrix}$.

Now, any Anosov diffeomorphism of M is homotopic to an affine diffeomorphism induced by an element $(\delta, \mathfrak{D}) \in \mathrm{Aff}(\mathbb{R}^n)$ where $\mathfrak{D}$ is hyperbolic and $(\delta, \mathfrak{D})$ normalizes E inside $\mathrm{Aff}(\mathbb{R}^n)$. As a consequence, $\mathfrak{D}$ must normalize the linear part of E inside $\mathrm{Gl}(n,\mathbb{R})$. This is however equivalent to $\mathfrak{D}$ commuting with A, or PDP^{-1} commuting with PAP^{-1}. It is easy to see that in this case $P\mathfrak{D}P^{-1}$ is of the form

$$P\mathfrak{D}P^{-1} = \begin{pmatrix} D_1 & 0 \\ 0 & D_2 \end{pmatrix}$$

where $D_1 \in \mathrm{Gl}(n-2,\mathbb{Q})$, $D_2 \in \mathrm{Gl}(2,\mathbb{Q})$. Both D_1 and D_2 are hyperbolic, because $\mathfrak{D}$ is hyperbolic. Moreover, as the characteristic polynomial of $\mathfrak{D}$ has integral coefficients and unit constant term, the characteristic polynomials of D_1 and D_2 also have integral coefficients and therefore also unit constant terms. In particular $\det(D_2) = \pm 1$.

Let us now make the following computations:

$$\det(I_n - \mathfrak{D}) = \det(I_n - P\mathfrak{D}P^{-1}) = \det(I_{n-2} - D_1)\det(I_2 - D_2)$$

$$\det(I_n - A\mathfrak{D}) = \det(I_n - PAP^{-1}P\mathfrak{D}P^{-1}) = \det(I_{n-2} - D_1)\det(I_2 + D_2)$$

Lemma 4.5 implies then that $\mathrm{sgn}(\det(I_n - \mathfrak{D})) \neq \mathrm{sgn}(\det(I_n - A\mathfrak{D}))$. (Notice that $\det(I_{n-2} - D_1) \neq 0$, since $\mathfrak{D}$ is hyperbolic.) Since the Nielsen and Lefschetz number are homotopic invariant we have that $N(f) \neq |L(f)|$.   □

Remark 4.11. *The condition that f is a Anosov diffeomorphism, is necessary. If f is for example the affine diffeomorphism on the manifold $M_{n,2}$ induced by* $\left(\begin{pmatrix} 0 \\ \vdots \\ 0 \end{pmatrix}, \begin{pmatrix} I_{n-2} & 0 \\ 0 & D_1 \end{pmatrix} \right)$ *with $D_1 \in \mathrm{Gl}(2,\mathbb{Z})$, then $\det(I_n - \mathfrak{D}) = \det(I_n - A\mathfrak{D}) = 0$. (For this f we have that $L(f) = N(f) = 0$.)*

4.3 Primitive flat manifolds

From now onwards, we concentrate on the class of primitive flat manifolds. These are the flat manifolds with $b_1(M) = 0$.

4.3.1 Primitive flat manifolds in dimension $n > 6$

We will work with the Bieberbach group $E_{k,l,m}$ which is generated by

$$(e_i, I_n) , (a, A) = \left(a, \begin{pmatrix} I_k & 0 & 0 \\ 0 & -I_l & 0 \\ 0 & 0 & -I_m \end{pmatrix}\right)$$

$$\text{and } (b, B) = \left(b, \begin{pmatrix} -I_k & 0 & 0 \\ 0 & I_l & 0 \\ 0 & 0 & -I_m \end{pmatrix}\right)$$

with $n = k + l + m$, 1 in the i-th place and 0 everywhere else for e_i $(1 \leq i \leq n)$, $a = (\frac{1}{2}\ 0 \cdots 0\ \frac{1}{2}\ 0 \cdots\ 0)^t$ and $b = (0 \cdots 0\ \frac{1}{2}\ 0 \cdots 0)^t$. (with $\frac{1}{2}$ on the first and $(k + l + 1)$-th place for a and $\frac{1}{2}$ on the $(k + 1)$-th place for b.) $E_{k,l,m}$ fits into the following short exact sequence:

$$1 \to \mathbb{Z}^n \to E_{k,l,m} \to \mathbb{Z}_2 \times \mathbb{Z}_2 \to 1$$

and $M_{k,l,m} = E_{k,l,m} \backslash \mathbb{R}^n$ is a flat manifold with $b_1(M_{k,l,m}) = 0$.

Remark 4.12. *The manifold $M_{k,l,m}$ used in this theorem is an arbitrary element of an interesting class of primitive $(\mathbb{Z}_2 \times \mathbb{Z}_2)$-manifolds introduced by P. Cobb in [10].*

Using analogous computations as in section 4.2.1 we can prove the following lemma:

Lemma 4.13. *Suppose $(\delta, \mathfrak{D}) \in \mathrm{Aff}(\mathbb{R}^n)$ and $n = k + l + m$. Then $(\delta, \mathfrak{D}) \in N_{\mathrm{Aff}(\mathbb{R}^n)}(E_{k,l,m})$ if and only if $(\delta, \mathfrak{D})$ is of one of the following forms:*

1.
$$\delta = \begin{pmatrix} x_1 \\ \vdots \\ x_n \end{pmatrix} \quad and \quad \mathfrak{D} = \begin{pmatrix} D_1 & 0 & 0 \\ 0 & D_2 & 0 \\ 0 & 0 & D_3 \end{pmatrix}$$

   with $x_1, \ldots, x_n \in \frac{1}{2}\mathbb{Z}$;

2.
$$\delta = \begin{pmatrix} x_1 \\ \vdots \\ x_n \end{pmatrix} \quad and \quad \mathfrak{D} = \begin{pmatrix} D_1 & 0 & 0 \\ 0 & 0 & D_2 \\ 0 & D_3 & 0 \end{pmatrix}$$

   with $x_1, x_{k+1}, x_{k+l+1} \in \frac{1}{4} + \frac{1}{2}\mathbb{Z}$ and the other $x_i \in \frac{1}{2}\mathbb{Z}$;

3.
$$\delta = \begin{pmatrix} x_1 \\ \vdots \\ x_n \end{pmatrix} \quad and \quad \mathfrak{D} = \begin{pmatrix} 0 & 0 & D_1 \\ 0 & D_2 & 0 \\ D_3 & 0 & 0 \end{pmatrix}$$

   with $x_{k+1} \in \frac{1}{4} + \frac{1}{2}\mathbb{Z}$ and the other $x_i \in \frac{1}{2}\mathbb{Z}$;

4.
$$\delta = \begin{pmatrix} x_1 \\ \vdots \\ x_n \end{pmatrix} \quad and \quad \mathfrak{D} = \begin{pmatrix} 0 & D_1 & 0 \\ 0 & 0 & D_2 \\ D_3 & 0 & 0 \end{pmatrix}$$

   with $x_1 \in \frac{1}{4} + \frac{1}{2}\mathbb{Z}$ and the other $x_i \in \frac{1}{2}\mathbb{Z}$;

5.
$$\delta = \begin{pmatrix} x_1 \\ \vdots \\ x_n \end{pmatrix} \quad and \quad \mathfrak{D} = \begin{pmatrix} 0 & 0 & D_1 \\ D_2 & 0 & 0 \\ 0 & D_3 & 0 \end{pmatrix}$$

with $x_1, x_{k+1} \in \frac{1}{4} + \frac{1}{2}\mathbb{Z}$ and the other $x_i \in \frac{1}{2}\mathbb{Z}$;

6.
$$\delta = \begin{pmatrix} x_1 \\ \vdots \\ x_n \end{pmatrix} \quad and \quad \mathfrak{D} = \begin{pmatrix} 0 & D_1 & 0 \\ D_2 & 0 & 0 \\ 0 & 0 & D_3 \end{pmatrix}$$

with $x_{k+l+1} \in \frac{1}{4} + \frac{1}{2}\mathbb{Z}$ and the other $x_i \in \frac{1}{2}\mathbb{Z}$.

where in each case $D_1 \in \mathrm{Gl}(k, \mathbb{Z})$, $D_2 \in \mathrm{Gl}(l, \mathbb{Z})$, $D_3 \in \mathrm{Gl}(m, \mathbb{Z})$ and for each D_i the first column is of the following form: the first entry is an odd integer, while the other entries are even.

With this lemma and Lemma 4.5 we can prove in a completely analogue way as Theorem 4.6 the following theorem:

Theorem 4.14. *For each dimension $n > 6$ there exists a primitive flat manifold M admitting Anosov diffeomorphisms f, g such that $N(f) \neq |L(f)|$ and $N(g) = |L(g)|$.*

Remark 4.15. *The theorem holds for every element $M_{k,l,m}$ of the class of $(\mathbb{Z}_2 \times \mathbb{Z}_2)$−manifolds introduced by Cobb which admits Anosov diffeomorphisms. Notice that it is an easy consequence of Theorem 4.3 that $M_{k,l,m}$ admits Anosov diffeomorphisms if and only if $k \geq 2, l \geq 2$ and $m \geq 2$ (see also [41]).*

4.3.2 Primitive flat manifolds in dimension 6

In this section we prove:

Theorem 4.16. *Let M be a 6-dimensional primitive flat manifold which admits Anosov diffeomorphisms. Then for each Anosov diffeomorphism $f : M \to M$:*

$$N(f) = |L(f)|.$$

C. Cid and T. Schulz present in [9] a complete enumeration of all six dimensional Bieberbach groups E with $b_1(E) = 0$. There are 5004 groups divided into 24 families and for each family they also describe

the $\mathbb{Q}$-decomposition of the holonomy representation. All the Bieberbach groups belonging to a given family have the same $\mathbb{Q}$- and $\mathbb{R}$-
decomposition. Therefore, we can use Theorem 4.3 to find the suitable
families containing Bieberbach groups which are fundamental groups
of flat manifolds admitting Anosov diffeomorphisms. Notice that each
one- and two-dimensional $\mathbb{Q}$-irreducible component of a representation
T is also $\mathbb{R}$-irreducible. Going over the list of all families, one finds that
there are only two suitable families (for the notation we refer to [9]):

1. **Family 1,1;1,1;1,1** (The corresponding holonomy representation
   decomposes into one-dimensional $\mathbb{Q}$−irreducible components, each
   with multiplicity 2.)
   In this family there are 4 Bieberbach groups E fitting into an exact
   sequence of the form

   $$1 \to \mathbb{Z}^6 \to E \to \mathbb{Z}_2 \times \mathbb{Z}_2 \to 1$$

   and they all belong to the same $\mathbb{Q}$-class. Note that the group $E_{2,2,2}$
   is one of the 4 Bieberbach groups belonging to this family.
2. **Family 2–1,2–1;1,1** (The corresponding holonomy representation
   decomposes into a one-dimensional and a two-dimensional $\mathbb{Q}$−irre-
   ducible component, each with multiplicity 2.)
   In this family there is one Bieberbach group E fitting into an exact
   sequence of the form

   $$1 \to \mathbb{Z}^6 \to E \to D_8 \to 1.$$

So in order to prove Theorem 4.16, we only have to deal with the 4
Bieberbach groups of the first family for which we treat the case of the
group $E_{2,2,2}$ in detail and the Bieberbach group of the second family
which we also treat in detail. The computations for the other 3 groups
are similar.

Proposition 4.17. *Suppose* $M_{2,2,2} = E_{2,2,2}\backslash\mathbb{R}^6$ *and* $f : M_{2,2,2} \to$
$M_{2,2,2}$ *is an Anosov diffeomorphism, then we have*

$$N(f) = |L(f)|.$$

<u>Proof:</u> Because of Lemma 4.2 and the fact that $L(f)$ and $N(f)$ are
homotopy invariants we may assume that f is a hyperbolic diffeomor-
phism induced by a $(\delta, \mathfrak{D}) \in N_{\text{Aff}(\mathbb{R}^6)}(E_{2,2,2})$. We can use Lemma 4.13

to give a full description of all possible lifts of hyperbolic affine diffeo-morphisms on $M_{2,2,2}$. We find that there are six types of hyperbolic diffeomorphisms. We will use Theorem 2.5 and so we have to calculate, for each of the six types of hyperbolic diffeomorphism, the determinants $\det(I_6 - \mathfrak{D})$, $\det(I_6 - A\mathfrak{D})$, $\det(I_6 - B\mathfrak{D})$ and $\det(I_6 - AB\mathfrak{D})$ and check that they have the same sign.

The numbering in the computations correspond to those of Lemma 4.13.

1. Then

$$\det(I_6 - \mathfrak{D}) = \det(I_2 - D_1)\det(I_2 - D_2)\det(I_2 - D_3)$$
$$\det(I_6 - A\mathfrak{D}) = \det(I_2 - D_1)\det(I_2 + D_2)\det(I_2 + D_3)$$
$$\det(I_6 - B\mathfrak{D}) = \det(I_2 + D_1)\det(I_2 - D_2)\det(I_2 + D_3)$$
$$\det(I_6 - AB\mathfrak{D}) = \det(I_2 + D_1)\det(I_2 + D_2)\det(I_2 - D_3)$$

Because of Lemma 4.5 we have that these determinants all have the same sign.

2. Suppose $K = d_{13}d_{13} + d_{23}d_{32} + d_{14}d_{41} + d_{24}d_{42}$, then we have

$$\det(I_6 - \mathfrak{D}) = \det(I_2 - D_1)(1 + \det(D_2)\det(D_3) - K)$$
$$= \det(I_6 - A\mathfrak{D}),$$
$$\det(I_6 - B\mathfrak{D}) = \det(I_2 + D_1)(1 + \det(D_2)\det(D_3) + K)$$
$$= \det(I_6 - AB\mathfrak{D})$$

and the eigenvalues of $\begin{pmatrix} 0 & D_2 \\ D_3 & 0 \end{pmatrix}$ are

$$\pm\sqrt{\frac{K \pm \sqrt{K^2 - 4\det(D_2)\det(D_3)}}{2}}.$$

Since D_2 and D_3 are elements of $\mathrm{Gl}(2,\mathbb{Z})$ we can distinguish two cases.

a) Either $\det(D_2)\det(D_3) = -1$, so $\det(I_6 - \mathfrak{D}) = -K\det(I_2 - D_1)$ and $\det(I_6 - B\mathfrak{D}) = K\det(I_2 + D_1)$. Thus because of Lemma 4.5 the signs of these two determinants must be the same.

b) Or $\det(D_2)\det(D_3) = 1$, so $\det(I_6 - \mathfrak{D}) = (2 - K)\det(I_2 - D_1)$ and $\det(I_6 - B\mathfrak{D}) = (2 + K)\det(I_2 + D_1)$. Thus because of Lemma 4.5 the signs of these two determinants are not equal if and only if $K = 0$ or $K = \pm 1$. But then the eigenvalues of $\mathfrak{D}$ are of absolute value one and this is not possible for a hyperbolic diffeomorphism.

3. and 6. Completely analogue to case 2, of course with another K.

4. and 5. In this case:

$$\det(I_6 - \mathfrak{D}) = \det(I_6 - A\mathfrak{D}) = \det(I_6 - B\mathfrak{D}) = \det(I_6 - AB\mathfrak{D}).$$

So for each type $(\delta, \mathfrak{D})$ we have that the four determinants have the same sign and because of Theorem 2.5 it follows that $N(f) = |L(f)|$ for each Anosov diffeomorphism f. $\square$

Remark 4.18. *The condition that f is a hyperbolic diffeomorphism is*

necessary. For example if f is induced by $(\begin{pmatrix} 0 \\ 0 \\ 0 \\ 0 \\ 0 \\ 0 \end{pmatrix}, \begin{pmatrix} 1\,1\,0\,0\,0\ \ 0 \\ 2\,1\,0\,0\,0\ \ 0 \\ 0\,0\,1\,1\,0\ \ 0 \\ 0\,0\,2\,1\,0\ \ 0 \\ 0\,0\,0\,0\,1\,-1 \\ 0\,0\,0\,0\,1\ \ 0 \end{pmatrix})$ *we*

have that $\det(I_6 - \mathfrak{D}) = 4$ *and* $\det(I_6 - A\mathfrak{D}) = -12$. *Hence* $L(f) = -4$ *and* $N(f) = 8$. *Use for instance Theorem 2.9 for the calculation of* $L(f)$ *and* $N(f)$.

Completely analogous computations can be done for the other three Bieberbach groups in family 1,1;1,1;1,1 For the single Bieberbach group in family 2–1,2–1;1,1 we have to work with the Bieberbach group E generated by

$$z_i = (e_i, I_6)\ ,\ (a, A) = (\begin{pmatrix} \frac{1}{2} \\ 0 \\ 0 \\ \frac{1}{2} \\ \frac{1}{4} \\ 0 \end{pmatrix}, \begin{pmatrix} 0 & 0 & 1 & 0 & 0 & 0 \\ 0 & 0 & 0 & 1 & 0 & 0 \\ -1 & 0 & 0 & 0 & 0 & 0 \\ 0 & -1 & 0 & 0 & 0 & 0 \\ 0 & 0 & 0 & 0 & 1 & 0 \\ 0 & 0 & 0 & 0 & 0 & 1 \end{pmatrix})$$

$$\text{and} \qquad (b, B) = (\begin{pmatrix} 0 \\ \frac{1}{2} \\ 0 \\ 0 \\ 0 \\ 0 \end{pmatrix}, \begin{pmatrix} 0 & 0 & 1 & 0 & 0 & 0 \\ 0 & 1 & 0 & 0 & 0 & 0 \\ 1 & 0 & 0 & 0 & 0 & 0 \\ 0 & 0 & 0 & -1 & 0 & 0 \\ 0 & 0 & 0 & 0 & -1 & 0 \\ 0 & 0 & 0 & 0 & 0 & -1 \end{pmatrix}).$$

In an analogue way as in the previous sections we first have to calculate $N_{\text{Aff}(\mathbb{R}^n)}(E)$:

Lemma 4.19. *Suppose* $(\delta, \mathfrak{D}) \in \mathrm{Aff}(\mathbb{R}^6)$, $D_1 \in \mathrm{Gl}(4, \mathbb{Z})$ *and* $D_2 \in \mathrm{Gl}(2, \mathbb{Z})$. *Then* $(\delta, \mathfrak{D}) \in N_{\mathrm{Aff}(\mathbb{R}^n)}(E)$ *if and only if* $\delta = (x_1, \ldots, x_6)$, $\mathfrak{D} = (d_{ij}) = \begin{pmatrix} D_1 & 0 \\ 0 & D_2 \end{pmatrix}$ *and the* x_i *and* d_{ij} *satisfies one of the following conditions:*

1. $x_1, x_2, x_3, x_4, x_6 \in \frac{1}{2}\mathbb{Z}$, $d_{56}, d_{66} \in \mathbb{Z}$, $d_{55} \in 1 + 4\mathbb{Z}$, $d_{65} \in 4\mathbb{Z}$,

   a) $x_5 \in \frac{1}{2}\mathbb{Z}$ and $D_1 = \begin{pmatrix} d_{11} & d_{12} & 0 & -d_{12} \\ d_{21} & d_{22} & d_{21} & 0 \\ 0 & d_{12} & d_{11} & d_{12} \\ -d_{21} & 0 & d_{21} & d_{22} \end{pmatrix}$

   with $d_{11}, d_{22} \in 1 + 2\mathbb{Z}$ and $d_{12}, d_{21} \in \mathbb{Z}$

   b) $x_5 \in \frac{1}{8} + \frac{1}{2}\mathbb{Z}$ and $D_1 = \begin{pmatrix} d_{11} & d_{12} & d_{11} & 0 \\ 0 & d_{22} & d_{23} & d_{22} \\ -d_{11} & 0 & d_{11} & d_{12} \\ -d_{23} & -d_{22} & 0 & d_{22} \end{pmatrix}$

   with $d_{12}, d_{23} \in 1 + 2\mathbb{Z}$ and $d_{11}, d_{22} \in \mathbb{Z}$

   c) $x_5 \in \frac{1}{4} + \frac{1}{2}\mathbb{Z}$ and $D_1 = \begin{pmatrix} 0 & d_{12} & d_{13} & d_{12} \\ d_{21} & 0 & -d_{21} & d_{24} \\ -d_{13} & -d_{12} & 0 & d_{12} \\ d_{21} & -d_{24} & d_{21} & 0 \end{pmatrix}$

   with $d_{13}, d_{24} \in 1 + 2\mathbb{Z}$ and $d_{12}, d_{21} \in \mathbb{Z}$

   d) $x_5 \in \frac{3}{8} + \frac{1}{2}\mathbb{Z}$ and $D_1 = \begin{pmatrix} d_{11} & 0 & -d_{11} & d_{14} \\ d_{21} & d_{22} & 0 & -d_{22} \\ d_{11} & -d_{14} & d_{11} & 0 \\ 0 & d_{22} & d_{21} & d_{22} \end{pmatrix}$

   with $d_{14}, d_{21} \in 1 + 2\mathbb{Z}$ and $d_{11}, d_{22} \in \mathbb{Z}$

2. $x_1, x_2, x_3, x_4, x_6 \in \frac{1}{2}\mathbb{Z}$, $d_{56}, d_{66} \in \mathbb{Z}$, $d_{55} \in 3 + 4\mathbb{Z}$, $d_{65} \in 4\mathbb{Z}$,

   a) $x_5 \in \frac{1}{2}\mathbb{Z}$ and $D_1 = \begin{pmatrix} 0 & d_{12} & d_{13} & d_{12} \\ d_{21} & d_{22} & d_{21} & 0 \\ d_{13} & d_{12} & 0 & -d_{12} \\ d_{21} & 0 & -d_{21} & -d_{22} \end{pmatrix}$

   with $d_{13}, d_{22} \in 1 + 2\mathbb{Z}$ and $d_{12}, d_{21} \in \mathbb{Z}$

   b) $x_5 \in \frac{1}{8} + \frac{1}{2}\mathbb{Z}$ and $D_1 = \begin{pmatrix} d_{11} & d_{12} & d_{11} & 0 \\ d_{21} & d_{22} & 0 & -d_{22} \\ d_{11} & 0 & -d_{11} & -d_{12} \\ 0 & -d_{22} & -d_{21} & -d_{22} \end{pmatrix}$

   with $d_{12}, d_{21} \in 1 + 2\mathbb{Z}$ and $d_{11}, d_{22} \in \mathbb{Z}$

$$c) \; x_5 \in \tfrac{1}{4} + \tfrac{1}{2}\mathbb{Z} \; and \; D_1 = \begin{pmatrix} d_{11} & d_{12} & 0 & -d_{12} \\ d_{21} & 0 & -d_{21} & d_{24} \\ 0 & -d_{12} & -d_{11} & -d_{12} \\ -d_{21} & d_{24} & -d_{21} & 0 \end{pmatrix}$$

with $d_{11}, d_{24} \in 1 + 2\mathbb{Z}$ and $d_{12}, d_{21} \in \mathbb{Z}$

$$d) \; x_5 \in \tfrac{3}{8} + \tfrac{1}{2}\mathbb{Z} \; and \; D_1 = \begin{pmatrix} d_{11} & 0 & -d_{11} & d_{14} \\ 0 & d_{22} & d_{23} & d_{22} \\ -d_{11} & d_{14} & -d_{11} & 0 \\ d_{23} & d_{22} & 0 & -d_{22} \end{pmatrix}$$

with $d_{14}, d_{23} \in 1 + 2\mathbb{Z}$ and $d_{11}, d_{22} \in \mathbb{Z}$

Now we can prove the following proposition:

Propositie 4.20. *Suppose* $M = E \backslash \mathbb{R}^6$ *and* $f : M \to M$ *is a hyperbolic diffeomorphism induced by a* $(\delta, \mathfrak{D}) \in \mathrm{Aff}(\mathbb{R}^6)$, *then we have*

$$N(f) = |L(f)|.$$

Proof: As in proposition 4.17 we will use theorem 2.5 to prove the statement and so we have to calculate the eight determinants for each type $(\delta, \mathfrak{D})$ of lemma 4.19. In an analogue way as in proposition 4.17 we can prove that the sign of the determinants is equal for hyperbolic $\mathfrak{D}$. (Again we find that if the signs are not equal, $\mathfrak{D}$ must have an eigenvalue of absolute value one.)   □

Remark 4.21. *The condition that* f *is a hyperbolic diffeomorphism, is necessary. For if* f *is induced by* $\left(\begin{pmatrix} 0 \\ 0 \\ 0 \\ 0 \\ 0 \\ 0 \end{pmatrix}, \begin{pmatrix} 1 & -1 & 0 & 1 & 0 & 0 \\ 1 & -1 & 1 & 0 & 0 & 0 \\ 0 & -1 & 1 & -1 & 0 & 0 \\ -1 & 0 & 1 & -1 & 0 & 0 \\ 0 & 0 & 0 & 0 & 1 & 1 \\ 0 & 0 & 0 & 0 & 4 & 3 \end{pmatrix} \right)$ *we have*

that $\det(I_6 - D) = -16$ *and* $\det(I_6 - BD) = 16$. *(For this* f *we have* $L(f) = 4$ *and* $N(f) = 12$.)

As an immediate consequence of Propositions 4.17 and 4.20, we conclude that Theorem 4.16 holds.

In the following table we summarize the results obtained in this chapter. Note that the $f : M \to M$ in this table stands for an Anosov diffeomorphism of the n-dimensional flat Riemannian manifolds M mentioned above and that we indicate if there exists an f which satisfies the relation.

$b_1(M)$	n	$N(f) = |L(f)|$	$N(f) \neq |L(f)|$	Proof
$b_1(M) = 0$	$n = 6$	always	never	Theorem 4.16
$b_1(M) = 0$	$n > 6$	exists	exists	Theorem 4.14
$2 \leq b_1(M) < n - 2$	$n > 4$	exists	exists	Theorem 4.6
$b_1(M) = n - 2$	$n \geq 4$	never	always	Theorem 4.10
$b_1(M) = n$	$n \geq 1$	always	never	see [1]

Chapter 5

Infra-nilmanifolds with cyclic holonomy group

In this chapter we examine the infra-nilmanifolds with cyclic holonomy group.[4] From the observations of Anosov we already know that in general the Anosov theorem does not hold for such manifolds. Indeed, the Klein bottle is a flat manifold with $\mathbb{Z}_2$ as holonomy group. However, in the second section we establish a sufficient condition for the Anosov relation to hold. More concretely, suppose that x_0 generates the holonomy group F of an infra-nilmanifold with cyclic holonomy group and $T : F \to \mathrm{Aut}(G)$ is the associated holonomy representation. Then we show that the Anosov theorem still holds in case -1 is not an eigenvalue of $T_*(x_0)$. So we extend the Anosov theorem to a new (and large class) of infra-nilmanifolds.

In the case of infra-nilmanifolds with odd order holonomy group of Chapter 3, -1 is never an eigenvalue of any of the matrices obtained from the holonomy representation, since these matrices are all of odd order. Therefore these manifolds also satisfy the above condition. However the situation is in general more delicate for the infra-nilmanifolds with cyclic holonomy group (of even order $2k$), since although if -1 is not an eigenvalue of $T_*(x_0)$, it can become an eigenvalue of some powers of $T_*(x_0)$.

In the third section we examine wether this condition is also necessary and we need to distinguish two cases. For flat manifolds with cyclic holonomy group we are able to show that the condition is also necessary and so for these manifolds the case is completely solved. On the other hand, in Chapter 3 we already gave an example of an infra-nilmanifold with holonomy group $\mathbb{Z}_2$ for which the Anosov theorem holds. One can

---

[4] The results of this chapter can also be found in [16].

easily verify that the generator of the holonomy group does not satisfy the condition mentioned above. This shows again that the validity of the Anosov theorem for infra-nilmanifolds is much more complicated then for flat manifolds.

5.1 Cyclic groups of matrices

As one might expect the results in this chapter depend heavily on the fact that we work with infra-nilmanifolds with cyclic holonomy group. To take advantage of this we need two lemmas concerning matrices of finite order (i.e. about cyclic groups of matrices). We leave the proof of these lemmas to the reader.

Lemma 5.1. *Let $B \in \mathrm{Gl}(n, \mathbb{R})$ be of order d and let $d_0, d_1, \ldots, d_t$ be the divisors of d such that $1 = d_0 < d_1 < \cdots < d_t = d$. Then there exists $n_0, n_1, \ldots, n_t \in \mathbb{N}$ and a $P \in \mathrm{Gl}(n, \mathbb{R})$ such that $n_0 + n_1 + \cdots + n_t = n$ and*

$$PBP^{-1} = \begin{pmatrix} B_0 & 0 & \cdots & 0 \\ 0 & B_1 & \cdots & 0 \\ \vdots & \vdots & \ddots & \vdots \\ 0 & 0 & \cdots & B_t \end{pmatrix}$$

with $B_i \in \mathrm{Gl}(n_i, \mathbb{R})$ and B_i only has eigenvalues of order exactly d_i $(0 \le i \le t)$.

Of course, by B being of order d, we mean that $B^d = I_n$, and an eigenvalue λ of order d_i indicates that $\lambda^{d_i} = 1$. Note that certain n_i from the lemma might be 0.

Lemma 5.2. *Let $d > 0$ be an integer with divisors $d_0, d_1, \ldots, d_t$ such that $1 = d_0 < d_1 < \cdots < d_t = d$. Suppose $B = \begin{pmatrix} B_0 & 0 & \cdots & 0 \\ 0 & B_1 & \cdots & 0 \\ \vdots & \vdots & \ddots & \vdots \\ 0 & 0 & \cdots & B_t \end{pmatrix}$ with*

$B_i \in \mathrm{Gl}(n_i, \mathbb{R})$ and B_i only has eigenvalues of order d_i $(0 \le i \le t)$. Let $n = n_0 + \cdots + n_t$ and suppose $C \in M_n(\mathbb{R})$ such that $CB = B^l C$ with $0 \le l < d$. Then

$$C = \begin{pmatrix} C_0 & 0 & \cdots & 0 \\ * & C_1 & \cdots & 0 \\ \vdots & \vdots & \ddots & \vdots \\ * & * & \cdots & C_t \end{pmatrix}$$

with $C_i \in M_{n_i}(\mathbb{R})$ and $*$ indicates any block of real numbers $(0 \leq i \leq t)$.

Now suppose that M is an infra-nilmanifold with a cyclic holonomy group F, generated by an element x_0 and $T : F \to \mathrm{Aut}(G)$ is the associated holonomy representation. Assume that F, and so x_0, is of order $2^r k$ with $r \geq 0$ and k an odd integer. Let $f : M \to M$ be a continuous map and $(\delta, \mathfrak{D})$ be a homotopy lift of f.

Because of Theorem 1.10 we know that there exists an integer l, with $0 \leq l < 2^r k$, such that $T_*(x_0^l)\mathfrak{D}_* = \mathfrak{D}_* T_*(x_0)$. Therefore we can apply the previous lemmas to $T_*(x_0)$ and $\mathfrak{D}_*$. Suppose $d_0, d_1, \ldots, d_t$ are the divisors of $2^r k$ and suppose $1 = d_0 < d_1 < \cdots < d_t = 2^r k$. Because of Lemma 5.1 there exists $n_0, n_1, \ldots, n_t \in \mathbb{N}$ and a $P \in \mathrm{Gl}(n, \mathbb{R})$ such that $n_0 + n_1 + \cdots + n_t = n$ and

$$PT_*(x_0)P^{-1} = \begin{pmatrix} A_0 & 0 & \cdots & 0 \\ 0 & A_1 & \cdots & 0 \\ \vdots & \vdots & \ddots & \vdots \\ 0 & 0 & \cdots & A_t \end{pmatrix}$$

with $A_i \in \mathrm{Gl}(n_i, \mathbb{R})$ and A_i only has eigenvalues of order d_i $(0 \leq i \leq t)$. Each d_i can be written as $2^s d$ with $s \geq 0$ and d an odd integer. Note that since $T_*(x_0)$ is of finite order, the only possible eigenvalues are ± 1 or non real eigenvalues with absolute value equal to one. Also note that any of the n_i can be zero.

Because of Lemma 5.2 we then have

$$P\mathfrak{D}_* P^{-1} = \begin{pmatrix} D_0 & 0 & \cdots & 0 \\ * & D_1 & \cdots & 0 \\ \vdots & \vdots & \ddots & \vdots \\ * & * & \cdots & D_t \end{pmatrix}$$

with $D_i \in M_{n_i}(\mathbb{R})$ $(0 \leq i \leq t)$. We will use this notations throughout this chapter.

Finally the following lemma will be very useful.

Lemma 5.3. Let $B, C \in M_n(\mathbb{R})$ be two real matrices such that $BC = CB$ and B has only non real eigenvalues. Then the multiplicity of any real eigenvalue of C must be even, which implies that $\det(I_n - C) \geq 0$.

Proof: We prove this lemma by induction on n. Note that n is even because B only has non real eigenvalues.

Suppose $n = 2$ and λ is a real eigenvalue of C with eigenvector

v such that $Cv = \lambda v$. Then Bv is also an eigenvector of C, since $CBv = BCv = \lambda Bv$. Moreover, v and Bv are linearly independent over $\mathbb{R}$. Otherwise there would exist a $\mu \in \mathbb{R}$ such that $Bv = \mu v$ contradicting the fact that B has no real eigenvalues. So the dimension of the eigenspace of λ is 2 and therefore the multiplicity of λ must be 2. Suppose the lemma holds for $r \times r$ matrices with r even and $r < n$. We then have to show that the lemma holds for $n \times n$ matrices. Again, let λ be a real eigenvalue of C and v an eigenvector of C such that $Cv = \lambda v$. Then, for any $m \in \mathbb{N}$, we have that $B^m v$ is an eigenvector of C. Indeed, $CB^m v = B^m Cv = \lambda B^m v$. Let S be the subspace of $\mathbb{R}^n$ generated by all vectors $B^m v$ with $m \in \mathbb{N}$. Then, for any $s \in S$, we have that $Cs = \lambda s$, so S is part of the eigenspace of λ and secondly $Bs \in S$, which implies that S is a B-invariant subspace of $\mathbb{R}^n$. Let $\{v_1, \ldots, v_k\}$ be a basis for S, then we can complete this basis with $v_{k+1}, \ldots, v_n$ to obtain a basis for $\mathbb{R}^n$. Writing (the matrices of the linear transformations determined by) B and C with respect to this new basis, implies the existence of a matrix $P \in \mathrm{Gl}(n, \mathbb{R})$ such that

$$PCP^{-1} = \begin{pmatrix} \lambda I_k & C_2 \\ 0 & C_3 \end{pmatrix} \quad \text{and} \quad PBP^{-1} = \begin{pmatrix} B_1 & B_2 \\ 0 & B_3 \end{pmatrix}$$

with B_1 a real $k \times k$ matrix; B_2, C_2 real $k \times (n-k)$ matrices; and B_3, C_3 real $(n-k) \times (n-k)$ matrices. Of course, the eigenvalues of B_1 and B_3 are also not real and $B_3 C_3 = C_3 B_3$. Therefore, k has to be even and we can proceed by induction on B_3 and C_3 to conclude that the real eigenvalues of C indeed have even multiplicities.

To prove the second claim of the lemma, we suppose that $\lambda_1, \ldots, \lambda_r$ are the real eigenvalues of C with even multiplicities $m_1, \ldots, m_r$ and that $\mu_1, \overline{\mu_1}, \ldots, \mu_t, \overline{\mu_t}$ are the complex eigenvalues of C with multiplicities $n_1, \ldots, n_t$. Then

$$\begin{aligned}
\det(I_n - C) &= (1 - \lambda_1)^{m_1} \cdots (1 - \lambda_r)^{m_r} \\
&\quad (1 - \mu_1)^{n_1}(1 - \overline{\mu_1})^{n_1} \cdots (1 - \mu_t)^{n_t}(1 - \overline{\mu_t})^{n_t} \\
&= (1 - \lambda_1)^{m_1} \cdots (1 - \lambda_r)^{m_r} \\
&\quad ((1 - \mu_1)(\overline{1 - \mu_1}))^{n_1} \cdots ((1 - \mu_t)(\overline{1 - \mu_t}))^{n_t} \\
&= (1 - \lambda_1)^{m_1} \cdots (1 - \lambda_r)^{m_r} |1 - \mu_1|^{2n_1} \cdots |1 - \mu_t|^{2n_t}
\end{aligned}$$

This last expression is clearly nonnegative since the m_i are even. $\square$

5.2 The Anosov theorem for infra-nilmanifolds with cyclic holonomy group

This section is completely devoted to the proof of the main result

Theorem 5.4. *Let M be an infra-nilmanifold with cyclic holonomy group F generated by x_0. Let $T : F \to \mathrm{Aut}(G)$ be the holonomy representation and suppose -1 is not an eigenvalue of $T_*(x_0)$. Then for any continuous map $f : M \to M$ we have that $N(f) = |L(f)|$.*

The proof of this theorem will be based on Theorem 2.5 and therefore we have to examine the sign of determinants $\det(I_n - T_*(x_0^m)\mathfrak{D}_*)$ for $0 \leq m < 2^r k$. Using the notations of the previous section we have that

$$\det(I_n - T_*(x_0^m)\mathfrak{D}_*) = \det(I_n - PT_*(x_0^m)P^{-1}P\mathfrak{D}_*P^{-1})$$
$$= \det(I_{n_0} - A_0^m D_0)\cdots\det(I_{n_t} - A_t^m D_t)$$

So it suffices to consider the determinants $\det(I_{n_i} - A_i^m D_i)$ separately. This allows us to reduce our investigation to the study of the sign of determinants of the form

$$\det(I_n - A^m D), \quad 0 \leq m < 2^s d$$

where

1. $2^s d | 2^r k$ with d an odd integer,
2. $A^{2^s d} = I_n$,
3. each eigenvalue of A is **exactly** of order $2^s d$ and
4. $DA = A^l D$ for some l, with $0 \leq l < 2^s d$.

We will distinguish several cases, depending on the possible values of s, d and l. Note that we will not need the case $s = 1$ and $d = 1$ (which corresponds to an eigenvalue -1 in $T_*(x_0)$).

In order to deal with all possible cases, we have to prove a series of lemmas. In fact, we are going to use the following scheme in our treatment:

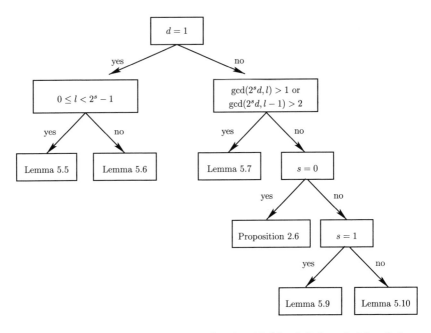

We start by looking at matrices of order 2^s (the left hand side of the scheme) and distinguish two cases depending on the value of l.

Lemma 5.5. *Suppose $A \in \mathrm{Gl}(n, \mathbb{R})$ only has eigenvalues of order 2^s with $s \geq 2$. Suppose $D \in M_n(\mathbb{R})$ such that $DA = A^l D$ with $0 \leq l < 2^s - 1$. Then the multiplicity of any real eigenvalue of D must be even and so we have that $\det(I_n - A^m D) \geq 0$ for any $0 \leq m < 2^s$.*

Proof: We prove this lemma by induction on s. Therefore we first look at the case where $s = 2$ and so l can be equal to $0, 1$ or 2. If $l = 0$ then $DA = D$ or $D(A - I_n) = 0_n$. Since 1 can not be an eigenvalue of A, we have that $A - I_n$ is invertible and so $D = 0_n$. Note that n is even since A only has non real eigenvalues and so in this case the lemma holds. If $l = 1$ we can apply Lemma 5.3 to A and D. Finally if $l = 2$ we have that $DA^2 = A^4 D = D$ and again $D = 0_n$ since $A^2 = -I_n$.

Assume now that $s > 2$ and that the lemma holds for smaller values of s. Since $DA = A^l D$ we also have that $DA^2 = (A^2)^l D$ and we can apply the induction hypothesis on A^2 which is of order 2^{s-1}. So we obtain for any l with $0 \leq l < 2^{s-1} - 1$ that the lemma holds, but the lemma holds already also for any l with $2^{s-1} \leq l < 2^s - 1$. Indeed for this l

we can consider $l' = l - 2^{s-1}$, then $0 \leq l' < 2^{s-1} - 1$ and we obtain $DA^2 = (A^2)^l D = (A^2)^{l'+2^{s-1}} D = (A^2)^{l'} D$. So, we can again apply the induction hypothesis.

There is still one case left, namely if $l = 2^{s-1} - 1$. Then we have that

$$DA^{2^{s-1}-1} = (A^{2^{s-1}-1})^{(2^{s-1}-1)}D = A^{(2^{s-1}-1)^2}D$$

Now $(2^{s-1} - 1)^2 = 2^{2s-2} - 2 \cdot 2^{s-1} + 1 = 2^s(2^{s-2} - 1) + 1$ and therefore $DA^{2^{s-1}-1} = AD$. This implies that $D(A + A^{2^{s-1}-1}) = (A + A^{2^{s-1}-1})D$. If we can show that $A + A^{2^{s-1}-1}$ only has non real eigenvalues then we can apply Lemma 5.3 to obtain that the lemma also holds in this case. Since A only has eigenvalues of order 2^s we know that $A + A^{2^{s-1}-1}$ only has eigenvalues of the form

$$e^{i \cdot \frac{2\pi t}{2^s}} + e^{i \cdot \frac{2\pi t}{2^s} \cdot (2^{s-1}-1)}$$

with $\gcd(t, 2^s) = 1$. Therefore t is odd and the imaginary part of such an eigenvalue is equal to

$$\sin(\frac{2\pi t}{2^s}) + \sin(\frac{2\pi t}{2^s}(2^{s-1} - 1)) = \sin(\frac{2\pi t}{2^s}) + \sin(\pi t - \frac{2\pi t}{2^s})$$
$$= \sin(\frac{2\pi t}{2^s}) + \sin(\pi - \frac{2\pi t}{2^s})$$
$$= 2\sin(\frac{2\pi t}{2^s})$$

Note that $\frac{2\pi t}{2^s}$ can not be equal to $0, \pi$ or 2π since $\gcd(t, 2^s) = 1$ and therefore the eigenvalues of $A + A^{2^{s-1}-1}$ are always non real.

As is shown is Lemma 5.3, this implies also the second statement of the lemma.   □

The second case is $l = 2^s - 1$ and there we obtain another result.

Lemma 5.6. *Suppose $A \in \mathrm{Gl}(n, \mathbb{R})$ only has eigenvalues of order 2^s with $s \geq 2$. Suppose $D \in M_n(\mathbb{R})$ such that $DA = A^{2^s-1}D$, then for any m, $0 \leq m < 2^s$,*

$$\det(I_n - D) = \det(I_n - A^m D)$$

<u>Proof:</u> Let $\mu_1, \overline{\mu_1}, \ldots \mu_w, \overline{\mu_w}$ be the different, non real eigenvalues of A with multiplicity $m_1, \ldots m_w$. A is diagonalizable so there exist a $Q \in \mathrm{Gl}(n, \mathbb{C})$ such that

$$QAQ^{-1} = \begin{pmatrix} \mu_1 I_{m_1} & 0 & \cdots & 0 & 0 \\ 0 & \overline{\mu_1} I_{m_1} & \cdots & 0 & 0 \\ \vdots & & \ddots & & \vdots \\ 0 & 0 & \cdots & \mu_w I_{m_w} & 0 \\ 0 & 0 & \cdots & 0 & \overline{\mu_w} I_{m_w} \end{pmatrix}$$

Since all the eigenvalues are of order 2^s this implies that

$$QA^{2^s-1}Q^{-1} = \begin{pmatrix} \overline{\mu_1} I_{m_1} & 0 & \cdots & 0 & 0 \\ 0 & \mu_1 I_{m_1} & \cdots & 0 & 0 \\ \vdots & & \ddots & & \vdots \\ 0 & 0 & \cdots & \overline{\mu_w} I_{m_w} & 0 \\ 0 & 0 & \cdots & 0 & \mu_w I_{m_w} \end{pmatrix}$$

One can easily verify that because of $DA = A^{2^{s-1}}D$ we obtain that

$$QDQ^{-1} = \begin{pmatrix} 0 & D_{i_1} & \cdots & 0 & 0 \\ D'_{i_1} & 0 & \cdots & 0 & 0 \\ \vdots & & \ddots & & \vdots \\ 0 & 0 & \cdots & 0 & D_{i_w} \\ 0 & 0 & \cdots & D'_{i_w} & 0 \end{pmatrix}$$

with $D_{i_j}, D'_{i_j} \in M_{m_j}(\mathbb{C})$, $1 \leq j \leq w$. Then since the eigenvalues of A_i are roots of unity we can calculate $\det(I_n - A^m D)$ for any m, $0 \leq m < 2^s$.

$$\det(I_n - A^m D)$$
$$= \det(I_n - QA^m Q^{-1} Q D Q^{-1})$$

$$= \det \begin{pmatrix} I_{m_1} & -(\mu_1)^m D_{i_1} & \cdots & 0 & 0 \\ -(\overline{\mu_1})^m D'_{i_1} & I_{m_1} & \cdots & 0 & 0 \\ \vdots & & \ddots & & \vdots \\ 0 & 0 & \cdots & I_{m_w} & -(\mu_w)^m D_{i_w} \\ 0 & 0 & \cdots & -(\overline{\mu_w})^m D'_{i_w} & I_{m_w} \end{pmatrix}$$

$$= \mu_1^{m\,m_1} \cdots \mu_w^{m\,m_w} \det \begin{pmatrix} \overline{\mu_1}^m I_{m_1} & -D_{i_1} & \cdots & 0 & 0 \\ -(\overline{\mu_1})^m D'_{i_1} & I_{m_1} & \cdots & 0 & 0 \\ \vdots & & \ddots & & \vdots \\ 0 & 0 & \cdots & \overline{\mu_w}^m I_{m_w} & -D_{i_w} \\ 0 & 0 & \cdots & -(\overline{\mu_w})^m D'_{i_w} & I_{m_w} \end{pmatrix}$$

$$= (\mu_1 \overline{\mu_1})^{m\,m_1} \cdots (\mu_w \overline{\mu_w})^{m\,m_w} \det \begin{pmatrix} I_{m_1} & -D_{i_1} & \cdots & 0 & 0 \\ -D'_{i_1} & I_{m_1} & \cdots & 0 & 0 \\ \vdots & & \ddots & & \vdots \\ 0 & 0 & \cdots & I_{m_w} & D_{i_w} \\ 0 & 0 & \cdots & D'_{i_w} & I_{m_w} \end{pmatrix}$$

$$= \det(I_n - D)$$

$\square$

From now on we consider matrices of order $2^s d$ with $d > 1$. Again we will distinguish several cases depending on the value of l. In a first case we consider l for which $\gcd(2^s d, l) > 1$ or $\gcd(2^s d, l-1) > 2$.

Lemma 5.7. Let $d > 1$ be an odd integer and let $s \geq 0$. Suppose $A \in \mathrm{Gl}(n, \mathbb{R})$ only has eigenvalues of order $2^s d$ and suppose $D \in M_n(\mathbb{R})$ such that $DA = A^l D$ with $0 \leq l < 2^s d$.
If $\gcd(2^s d, l) > 1$ or $\gcd(2^s d, l-1) > 2$ then we have that $\det(I_n - A^m D) \geq 0$ for any m.

Proof: Assume first that $\gcd(2^s d, l) = l_1 > 1$ and that $l = l_1 l_2$ and $2^s d = l_1 d'$. Then we have

$$DA^{d'} = (A^{d'})^l D$$
$$= A^{2^s d l_2} D$$
$$= D$$

Now 1 is not an eigenvalue of $A^{d'}$ since $d' < 2^s d$ and the eigenvalues of A are of order $2^s d$. This implies as before that $D = 0_n$ and $\det(I_n - A^m D) = 1 \geq 0$ for any m.

Secondly, assume that $\gcd(2^s d, l-1) = l_1 > 2$ and that $l-1 = l_1 l_2$ and $2^s d = l_1 d'$. Then we have that

$$
\begin{aligned}
DA^{d'} &= (A^{d'})^l D \\
&= A^{d'(1+l_1 l_2)} D \\
&= A^{d'+2^s d l_2} D \\
&= A^{d'} D
\end{aligned}
$$

Now $A^{d'}$ only has non real eigenvalues since $d' < 2^{s-1}d$ and the eigenvalues of A are of order $2^s d$. So we can apply Lemma 5.3 to finish the proof of this lemma. $\qquad\square$

Finally, we are led to the situation in which we have to consider those l for which $\gcd(2^s d, l) = 1$ and $\gcd(2^s d, l-1) \leq 2$. If $s = 0$, then we can apply Theorem 3.3 since d is odd. On the other hand if $s \geq 1$, then $\gcd(2^s d, l-1) = 2$, since $\gcd(2^s d, l) = 1$ implies that l is odd.
The following lemma is useful to solve the second case.

Lemma 5.8. *Let $d > 1$ be an odd integer and let $s \geq 1$. Suppose $A \in \mathrm{Gl}(n, \mathbb{R})$ only has eigenvalues of order $2^s d$ and suppose $D \in M_n(\mathbb{R})$ such that $DA = A^l D$ with $0 \leq l < 2^s d$.*
If $\gcd(2^s d, l-1) = 2$ then we have that $\det(I_n - D) = \det(I_n - A^{2m} D)$ and $\det(I_n - AD) = \det(I_n - A^{2m+1} D)$ for any m.

Proof: We can establish for any m the following relations between the determinants.

$$
\begin{aligned}
\det(I_n - D) &= \det(A^m - DA^m)\det(A^{-m}) \\
&= \det(A^{-m})\det(A^m - A^{ml}D) \\
&= \det(I_n - A^{m(l-1)}D)
\end{aligned}
$$

Since $\gcd(l-1, 2^s d) = 2$, we have that the group generated by A^{l-1} is of order $2^{s-1}d$ and thus consists of all even powers of A. It follows that $\det(I_n - D) = \det(I_n - A^{2m}D)$ for any m.
The second part can be proved analogously with $D' = AD$. $\qquad\square$

With this last lemma, we can prove the following lemma for $s = 1$.

Lemma 5.9. *Let $d > 1$ be an odd integer. Suppose $A \in \mathrm{Gl}(n, \mathbb{R})$ only has eigenvalues of order $2d$ and suppose $D \in M_n(\mathbb{R})$ such that $DA = A^l D$ with $0 \leq l < 2d$.*
If $\gcd(2d, l) = 1$ and $\gcd(2d, l-1) = 2$, then $\det(I_n - A^m D_i) \geq 0$ for any m or $\det(I_n - A^m D_i) \leq 0$ for any m.

Proof: Because of Lemma 5.8 we only have to prove the lemma for $\det(I_n - D)$ and $\det(I_n - AD)$. Moreover since d is odd and the eigenvalues of A^d are of order 2, it suffices to prove the lemma for $\det(I_n - D)$ and $\det(I_n - AD) = \det(I_n - A^d D) = \det(I_n + D)$.

Since $\gcd(2d, l) = 1$, Euler's theorem tells us that $l^{\phi(2d)} \equiv 1 \bmod 2d$. This implies that

$$\begin{aligned}
D^{\phi(2d)} A &= D^{\phi(2d)-1} A^l D \\
&= D^{\phi(2d)-2} A^{l^2} D^2 \\
&= \ldots \\
&= A^{l^{\phi(2d)}} D^{\phi(2d)} \\
&= A D^{\phi(2d)}
\end{aligned}$$

So we obtain that, because of Lemma 5.3, the multiplicity of the real eigenvalues of $D^{\phi(2d)}$ must be even.

Suppose that $\lambda_1, \ldots, \lambda_v$ are the real, positive eigenvalues of $D^{\phi(2d)}$ each with even multiplicity $m_1, \ldots, m_v$; that $\gamma_1, \ldots, \gamma_{t_1}$ are the real, strictly negative eigenvalues of $D^{\phi(2d)}$ and that $\mu_1, \overline{\mu_1}, \ldots, \mu_{t_2}, \overline{\mu_{t_2}}$ are the non real eigenvalues of $D^{\phi(2d)}$. The eigenvalues of D must be $\phi(2d)$-th roots of these eigenvalues of $D^{\phi(2d)}$. Note that $\phi(2d)$ is an even integer since $d > 1$ and therefore a $\phi(2d)$-th root of γ_i or μ_k is always non real. While the $\phi(2d)$-th root of the λ_j can be real or non real , say α_j. But if α_j is an eigenvalue of D, then $\overline{\alpha_j}$ also has to be an eigenvalue of D. Now $(\overline{\alpha_j})^{\phi(2d)} = \overline{\alpha_j^{\phi(2d)}} = \lambda_j$. So $\overline{\alpha_j}$ has to be a $\phi(2d)$-th root of the same λ_j. Since the m_j are even, this implies for each j, $1 \leq j \leq v$, that the number of real eigenvalues of D coming from λ_j must be even.

Let us for each j, $1 \leq j \leq v$, denote the positive real $\phi(2d)$-th root by δ_j (and the negative real root by $-\delta_j$). We denote the multiplicity of δ_j, resp. $-\delta_j$, as an eigenvalue of D by r_j, resp. s_j. It is of course possible that δ_j or $-\delta_j$ is not an eigenvalue of D. In this case we take its multiplicity to be equal to 0. We then always have that $r_j + s_j \in 2\mathbb{Z}$.

Using the same arguments as in the proof of Lemma 5.3 and using the above information we know that the only factors that matter are

$$(1 - \delta_1)^{r_1} \cdots (1 - \delta_v)^{r_v} (1 + \delta_1)^{s_1} \cdots (1 + \delta_v)^{s_v}$$

in $\det(I_n - D)$ and

$$(1 + \delta_1)^{r_1} \cdots (1 + \delta_v)^{r_v} (1 - \delta_1)^{s_1} \cdots (1 - \delta_v)^{s_v}$$

in $\det(I_n + D)$.

For each $i \in \{1, \ldots, v\}$ there is in $\det(I_n - D)$ a factor of the form $(1 - \delta_i)^{r_i}(1 + \delta_i)^{s_i}$ and in $\det(I_n + D)$ there is a factor of the form $(1 + \delta_i)^{r_i}(1 - \delta_i)^{s_i}$. Suppose that $r_i \geq s_i$ (the other case is completely similar). Then

$$(1 - \delta_i)^{r_i}(1 + \delta_i)^{s_i} = (1 - \delta_i)^{(r_i - s_i)}(1 - \delta_i)^{s_i}(1 + \delta_i)^{s_i}$$
$$= (1 - \delta_i)^{r_i - s_i}(1 - \delta_i^2)^{s_i}$$

and

$$(1 + \delta_i)^{r_i}(1 - \delta_i)^{s_i} = (1 + \delta_i)^{r_i - s_i}(1 - \delta_i^2)^{s_i}$$

Since $r_i + s_i \in 2\mathbb{Z}$, we have that $r_i - s_i$ is also an even integer. So in both cases the first factor is positive and the second factor is the same. This ends the proof of this lemma. $\square$

Finally we can prove the following lemma for $s \geq 2$.

Lemma 5.10. *Let $d > 1$ be an odd integer and $s \geq 2$. Suppose $A \in$ $\mathrm{Gl}(n, \mathbb{R})$ only has eigenvalues of order $2^s d$ and suppose $D \in M_n(\mathbb{R})$ such that $DA = A^l D$ with $0 \leq l < 2^s d$.*
If $\gcd(2^s d, l - 1) = 2$ and $l \not\equiv 2^s - 1 \bmod 2^s$, then $\det(I_n - A^m D) \geq 0$ for any m.
If $\gcd(2^s d, l - 1) = 2$ and $l \equiv 2^s - 1 \bmod 2^s$, then $\det(I_n - D) = \det(I_n - A^m D)$ for any m.

Proof: Because of Lemma 5.8 we only have to prove the lemma for $\det(I_n - D)$ and $\det(I_n - AD)$. Lemma 5.8 also implies that $\det(I_n - AD) = \det(I_n - A^d D)$ since d is odd. Now A^d only has eigenvalues of order 2^s and if $l \not\equiv 2^s - 1 \bmod 2^s$ we can apply Lemma 5.5 to D and A^d. If on the other hand $l \equiv 2^s - 1 \bmod 2^s$ then we can apply Lemma 5.6 to D and A^d. $\square$

With all these lemmas we can now prove the main result.

Proof of Theorem 5.4: Denote the order of F by $2^r k$ with k an odd integer. Let $(\delta, \mathfrak{D})$ be a homotopy lift of f and suppose that $\mathfrak{D}_* T_*(x_0) = T_*(x_0^l)\mathfrak{D}_*$. Suppose $d_0, d_1, \ldots, d_t$ are the divisors of $2^r k$ and suppose $1 = d_0 < d_1 < \cdots < d_t = 2^r k$. Because of Lemma 5.1 and the condition on $T_*(x_0)$ there exists $n_0, n_2, \ldots, n_t \in \mathbb{N}$ and a $P \in \mathrm{Gl}(n, \mathbb{R})$ such that $n_0 + n_2 + \cdots + n_t = n$ and

$$PT_*(x_0)P^{-1} = \begin{pmatrix} A_0 & 0 & \cdots & 0 \\ 0 & A_2 & \cdots & 0 \\ \vdots & \vdots & \ddots & \vdots \\ 0 & 0 & \cdots & A_t \end{pmatrix}$$

with $A_i \in \mathrm{Gl}(n_i, \mathbb{R})$ and A_i only has eigenvalues of order d_i. $(0 \leq i \leq t)$. Note that $n_1 = 0$ since -1 is not an eigenvalue of $T_*(x_0)$. Because of Lemma 5.2 we also have

$$P\mathfrak{D}_*P^{-1} = \begin{pmatrix} D_0 & 0 & \cdots & 0 \\ * & D_2 & \cdots & 0 \\ \vdots & \vdots & \ddots & \vdots \\ * & * & \cdots & D_t \end{pmatrix}$$

with $D_i \in M_{n_i}(\mathbb{R})$ $(0 \leq i \leq t)$ and $*$ can be any block of real numbers. If we want to use Theorem 2.5, we have to calculate

$$\det(I_n - T_*(x_0^m)\mathfrak{D}_*) = \det(I_{n_0} - A_0^m D_0) \cdots \det(I_{n_t} - A_t^m D_t)$$

for any m, $0 \leq m < 2^r k$. As explained before, we consider all the factors above separately (so we fix an i and see what happens when m varies). If $d_i = 2^s$ $(s \geq 2)$ then we can, depending on l, apply Lemma 5.5 or Lemma 5.6 to show that all these factors have the same sign. Note that in case s is zero $(i = 0)$, $\det(I_{n_0} - A_0^m D_0)$ does not depend on m. By our assumption, the case $s = 1$ does not occur.

If d_i is not a power of 2 then we can, again depending on the value of l (see the scheme on page 67 and the discussion immediately before Lemma 5.8), apply Lemma 5.7, Theorem 3.3, Lemma 5.9 or Lemma 5.10 to show that these factors also have the same sign.

So in each case we obtain that the condition in Theorem 2.5 is satisfied and for each f we have $N(f) = |L(f)|$. $\square$

Remark 5.11.

- *The condition that -1 is not an eigenvalue of $T_*(x_0)$ is crucial since D. Anosov constructed a counterexample on the Klein Bottle which has $\mathbb{Z}_2$ as holonomy group. In the following section we go deeper into this.*

- *In the following chapter we prove that the Anosov theorem holds for orientable generalized Hantzsche-Wendt manifolds. This implies that it is not straight forward to generalize Theorem 5.4 to infra-nilmanifolds with other holonomy groups if -1 is an eigenvalue.*

In the following sections we examine the infra-nilmanifolds M with cyclic holonomy group, for which the holonomy representation does not satisfy the condition of Theorem 5.4. We have to distinguish two cases for these manifolds. In the case that M is a flat manifold, we are always able to construct a continuous map $f : M \to M$ such that $N(f) \neq |L(f)|$. So in the case of flat manifolds with cyclic holonomy group, we have a complete picture.

As already mentioned in the introduction, the example in section 3.3 shows that the same does not hold for infra-nilmanifolds in general.

5.3 The sharpness of the main result for flat manifolds

In order to construct a continuous map f on a flat manifold with cyclic holonomy group which does not satisfy the condition in Theorem 5.4, we already know that we only have to work with orientable manifolds. Indeed, flat non-orientable manifolds always admit an expanding map f and so Theorem 3.6 implies that we have a counter example.

Note that, because of Proposition 1.6, the manifolds which satisfy the conditions of Theorem 5.4 are orientable.

For flat orientable manifolds we can prove the following proposition.

Proposition 5.12. *Let M be a n-dimensional, orientable, flat manifold with cyclic holonomy group F generated by x_0. Let $T : F \to \mathrm{Gl}(n, \mathbb{Z})$ be the associated holonomy representation and suppose -1 is an eigenvalue of $T_*(x_0)$.*
Then there always exists a continuous map $f : M \to M$ such that $N(f) \neq |L(f)|$.

Proof: As -1 is an eigenvalue of the generator of the holonomy representation, F has to be a cyclic group of even order, say $2m$.
$\pi_1(M)$ is a n-dimensional Bieberbach group with translation subgroup $Z \cong \mathbb{Z}^n$ and holonomy group $\mathbb{Z}_{2m} = \langle x_0 \rangle$, for some $m \geq 1$. $\pi_1(M)$ fits in a short exact sequence

$$1 \to Z \cong \mathbb{Z}^n \to \pi_1(M) \to \mathbb{Z}_{2m} = \langle x_0 \rangle \to 1. \qquad (5.1)$$

This short exact sequence determines a faithful representation $\varphi : \mathbb{Z}_{2m} \to \mathrm{Aut}(Z)$ (when viewed as a real representation is actually the same as the holonomy representation T). With respect to a good choice of generators of the free abelian group Z, φ is represented by blocked diagonal matrices, with

$$\varphi(x_0) = \begin{pmatrix} A(x_0) & 0 \\ * & C(x_0) \end{pmatrix}$$

where $A(x_0)$ only has eigenvalues ± 1 and $C(x_0)$ has no real eigenvalues. It follows that $A(x_0)$ is a matrix of order 2 and by eventually changing our set of generators for Z again, we can assume that

$$\varphi(x_0) = \begin{pmatrix} -I_s & 0 & 0 \\ * & I_t & 0 \\ * & * & C(x_0) \end{pmatrix}.$$

for some integers $s, t \geq 0$.

Now, -1 is an eigenvalue of $\varphi(x_0)$ ($s \neq 0$) and we also assume that M is orientable, which means that s is even, thus at least 2. Therefore, we will write

$$\varphi(x_0) = \begin{pmatrix} -1 & 0 & 0 & 0 & 0 \\ 0 & -1 & 0 & 0 & 0 \\ 0 & 0 & -I_{s-2} & 0 & 0 \\ * & * & * & I_t & 0 \\ * & * & * & * & C(x_0) \end{pmatrix}.$$

The group $\pi_1(M)$ is determined by a 2-cocycle $f : \mathbb{Z}_{2m} \times \mathbb{Z}_{2m} \to \mathbb{Z}^n$. This means that the group $\pi_1(M) = \mathbb{Z}^n \times \mathbb{Z}_{2m}$ (as a set) and the product in $\pi_1(M)$ is given by

$$\forall z, z' \in \mathbb{Z}^n, \ \forall x, y \in \mathbb{Z}_{2m} : \ (z, x)(z', y) = (z + \varphi(x)z' + f(x, y), xy).$$

Any element of $H^2(\mathbb{Z}_{2m}, \mathbb{Z}^n)$ has an order dividing $2m$. Therefore, there exists a map $g : \mathbb{Z}_{2m} \to \mathbb{Z}^n$, with $\delta g = 2mf$. (Recall that this means that $\delta g(x, y) = \varphi(x)g(y) - g(xy) + g(x) = 2mf(x, y)$)
It is now easy to check that

$$\psi_1 : \pi_1(M) = \mathbb{Z}^n \times \mathbb{Z}_{2m} \to \mathrm{Aff}(\mathbb{R}^n) : (z, x) \mapsto (z + \frac{g(x)}{2m}, \varphi(x))$$

realizes the group $\pi_1(M)$ as an affine group, with its translation subgroup Z mapped isomorphically onto $\mathbb{Z}^n$. Let us consider the image of $(0, x_0)$:

$$\psi_1(0, x_0) = \left(\left(\begin{array}{c} \frac{x}{2m} \\ \frac{y}{2m} \\ \frac{u_1}{2m} \\ \frac{u_2}{2m} \\ \frac{u_3}{2m} \end{array} \right), \left(\begin{array}{ccccc} -1 & 0 & 0 & 0 & 0 \\ 0 & -1 & 0 & 0 & 0 \\ 0 & 0 & -I_{s-2} & 0 & 0 \\ * & * & * & I_t & 0 \\ * & * & * & * & C(x_0) \end{array} \right) \right).$$

(For some $x, y \in \mathbb{Z}$, $u_1 \in \mathbb{Z}^{s-2}$, $u_2 \in \mathbb{Z}^t$, $u_3 \in \mathbb{Z}^{n-s-t}$).

Let $v = (-\frac{x}{4m}, -\frac{y}{4m}, 0, 0, 0)^t \in \mathbb{R}^n$ and take $\psi_2 = (v, I_n)\psi_1(v, I_n)^{-1}$.

Then, $\psi_2(z, 1) = \psi_1(z, 1)$ for all $z \in \mathbb{Z}^n$ and

$$\psi_2(0, x_0) = \left(\left(\begin{array}{c} 0 \\ 0 \\ \frac{u_1}{2m} \\ \frac{u_2}{2m} \\ \frac{u_3}{2m} \end{array} \right), \left(\begin{array}{ccccc} -1 & 0 & 0 & 0 & 0 \\ 0 & -1 & 0 & 0 & 0 \\ 0 & 0 & -I_{s-2} & 0 & 0 \\ * & * & * & I_t & 0 \\ * & * & * & * & C(x_0) \end{array} \right) \right).$$

There exists a rational matrix

$$P = \left(\begin{array}{ccccc} 1 & 0 & 0 & 0 & 0 \\ 0 & 1 & 0 & 0 & 0 \\ 0 & 0 & I_s & 0 & 0 \\ * & * & * & I_t & 0 \\ * & * & * & 0 & I_{n-s-t} \end{array} \right) \in \mathrm{Gl}(n, \mathbb{Q})$$

such that

$$P\varphi(x_0)P^{-1} = \left(\begin{array}{ccccc} -1 & 0 & 0 & 0 & 0 \\ 0 & -1 & 0 & 0 & 0 \\ 0 & 0 & -I_{s-2} & 0 & 0 \\ 0 & 0 & 0 & I_t & 0 \\ 0 & 0 & 0 & * & C(x_0) \end{array} \right).$$

Let D_1 be the matrix

$$D_1 = \left(\begin{array}{ccccc} 2 & 1 & 0 & 0 & 0 \\ 1 & 1 & 0 & 0 & 0 \\ 0 & 0 & I_{s-2} & 0 & 0 \\ 0 & 0 & 0 & I_t & 0 \\ 0 & 0 & 0 & 0 & I_{n-s-t} \end{array} \right).$$

It is obvious that D_1 commutes with $P\varphi(x_0)P^{-1}$ and thus $D_2 = P^{-1}D_1 P \in \mathrm{Gl}(n, \mathbb{Q})$ commutes with $\varphi(x_0)$. The matrix D_2 is an invertible rational matrix, whose characteristic polynomial (which is the same as the characteristic polynomial of D_1) has integer coefficients and unit constant term. This implies, by a result of H. Porteous ([49]), that there exists a positive integer k such that $D_3 = D_2^k \in \mathrm{Gl}(n, \mathbb{Z})$. Now, D_3 has almost all eigenvalues equal to 1, except two positive real eigenvalues, say $\lambda_1 > 1$ and $\lambda_2 = \frac{1}{\lambda_1} < 1$. Again by taking a suitable power of D_3, we obtain a new matrix $D_4 = D_3^l$ (for some l) such that its two eigenvalues different from 1 are $\lambda_1^l, \frac{1}{\lambda_1^l}$ and satisfy

$$\lambda_1^l > 2m + 1 \Rightarrow \frac{1}{\lambda_1^l} < \frac{1}{2m + 1}.$$

Now, finally, let $\mathfrak{D} = (2m + 1)D_4$. It is obvious that $\mathfrak{D}$ still commutes with $\varphi(x_0)$. The matrix $\mathfrak{D}$ is of the form

$$\mathfrak{D} = \begin{pmatrix} a\ b & 0 & 0 & 0 \\ b\ c & 0 & 0 & 0 \\ 0\ 0 & (2m+1)I_{s-2} & 0 & 0 \\ *\ * & * & (2m+1)I_t & 0 \\ *\ * & * & 0 & (2m+1)I_{n-s-t} \end{pmatrix}.$$

Each $*$ indicates a block with entries in $(2m+1)\mathbb{Z}$ and the block $\begin{pmatrix} a & b \\ b & c \end{pmatrix}$ has two positive real eigenvalues μ_1 and μ_2 satisfying: $\mu_1 > (2m + 1)^2$ and $\mu_2 < 1$.

Now, conjugating with $(0, \mathfrak{D})$ inside $\mathrm{Aff}(\mathbb{R}^n)$ induces an endomorphism of $\psi_2(E)$. This follows from the following observations:

1. Let $(z, I_n) \in \psi_2(E)$ (so $z \in \mathbb{Z}^n$), $(0, \mathfrak{D})(z, I_n)(0, \mathfrak{D}^{-1}) = (\mathfrak{D}z, I_n) \in \psi_2(E)$.
2. We compute the image of $\psi_2(0, x_0) = (t(x_0), \varphi(x_0))$, with $t(x_0) = (0, 0, \frac{u_1}{2m}, \frac{u_2}{2m}, \frac{u_3}{2m})^t$:

$$\begin{aligned} (0, \mathfrak{D})\psi_2(0, x_0) &= (0, \mathfrak{D})(t(x_0), \varphi(x_0)) \\ &= (\mathfrak{D}t(x_0), \mathfrak{D}\varphi(x_0)) \\ &= ((\mathfrak{D} - I_n)t(x_0), I_n)(t(x_0), \varphi(x_0))(0, \mathfrak{D}) \end{aligned}$$

By the construction of $\mathfrak{D}$ and the fact that the first 2 entries of $t(x_0)$ are zero, we have that $(\mathfrak{D} - I_n)t(x_0) \in \mathbb{Z}^n$. This implies that $(0, \mathfrak{D})(t(x_0), \varphi(x_0))(0, \mathfrak{D})^{-1} \in \psi_2(E)$.

Let f be the map on M, induced by conjugation with $(0, \mathfrak{D})$. On the one hand, we have that

$$\det(I_n - \mathfrak{D}) = \underbrace{(1 - \mu_1)}_{<0}\underbrace{(1 - \mu_2)}_{>0}\underbrace{(-2m)^{s-2}}_{>0}(-2m)^t\underbrace{(-2m)^{n-s-t}}_{>0},$$

while on the other hand

$$\det(I_n - \varphi(x_0)\mathfrak{D})$$
$$= \underbrace{(1 + \mu_1)}_{>0}\underbrace{(1 + \mu_2)}_{>0}\underbrace{(2 + 2m)^{s-2}}_{>0}(-2m)^t\underbrace{\det(I_{n-s-t} - (2m + 1)C(x_0))}_{>0}.$$

It is obvious that these two determinants differ in sign and Theorem 2.5 again implies that $N(f) \neq |L(f)|$.

□

Chapter 6

Generalized Hantzsche-Wendt manifolds

In this chapter we work with a special, but still large, class of flat manifolds, namely the generalized Hantzsche-Wendt manifolds. These are n-dimensional flat manifolds with holonomy group isomorphic to $\mathbb{Z}_2^{n-1}$. In dimension 2 the Klein bottle is the only example of such a manifold. In dimension 3 there are three examples, namely the classical Hantzsche-Wendt manifold, which is an orientable manifold, and two non-orientable manifolds. For a more detailed study of generalized Hantzsche-Wendt manifolds we refer to [26],[46], [47] and [51].

As shown in Chapter 3 we only have to concentrate on flat orientable generalized Hantzsche-Wendt manifolds, since for the non-orientable ones we already obtained that the Anosov theorem does not hold. Although we have a partial picture, there is still a large part missing since in [46] it is shown that the number of orientable Hantzsche-Wendt manifolds grows exponentially with the dimension.

Perhaps surprisingly, we are able to show that the Anosov theorem always holds in the class of orientable generalized Hantzsche-Wendt manifolds. [5] These manifolds are however 'totally' different from the infra-nilmanifolds discussed in the previous chapters. In general terms one can say that up till now we preferred the 'absence' of the eigenvalue -1 in the matrices obtained from the holonomy representation. Now we have the complete opposite, since in the first section we will show that the image of the corresponding holonomy representation, contains n matrices having -1 as an eigenvalue with (maximal!) multiplicity $n-1$.

---

[5] The results of this chapter can also be found in [13]

6.1 Definition and properties

As already mentioned, a n-dimensional flat manifold is called a generalized Hantzsche-Wendt (GHW) manifold if its holonomy group is isomorphic to $\mathbb{Z}_2^{n-1}$. In this case the Bieberbach group $E - \pi_1(M)$ is called a GHW group. A n-dimensional Bieberbach group E is said to be diagonal if its lattice $\mathbb{Z}^n$ of translations has an orthogonal basis for which the holonomy representation T diagonalizes. In [51] J.P. Rossetti and A. Szczepański proved the following theorem:

Theorem 6.1. *The fundamental group of a generalized Hantzsche-Wendt manifold is diagonal.*

Suppose M is a n-dimensional flat GHW manifold and $T : \mathbb{Z}_2^{n-1} \to \mathrm{Gl}(n, \mathbb{Z})$ is the associated holonomy representation. Because of Theorem 6.1 we may assume that $T(x)$ is diagonal for each $x \in \mathbb{Z}_2^{n-1}$ and hence we know that the diagonal elements must be 1 or -1.
If moreover, M is an orientable manifold, then for each $x \in \mathbb{Z}_2^{n-1}$, the diagonal entries of $T(x)$ consist of an even number of -1's while the others are 1. In fact it is obvious that in $\mathrm{Gl}(n, \mathbb{Z})$ there are exactly 2^{n-1} diagonal matrices whose diagonal entries consist of an even number of -1's while the other entries are 1. It follows that

Corollary 6.2. *Let M be a n-dimensional orientable flat GHW manifold and $T : \mathbb{Z}_2^{n-1} \to \mathrm{Gl}(n, \mathbb{Z})$ its associated holonomy representation. Then*

1. The image of $T : \mathbb{Z}_2^{n-1} \to \mathrm{Gl}(n, \mathbb{Z})$ is completely determined;
2. n is an odd integer;
3. the first Betti number of M is 0.

Proof: Note that, because of Theorem 6.1, T is a diagonal representation. Since a holonomy representation must be faithful and the holonomy group is of order 2^{n-1}, the first result follows immediately.
Suppose n is an even integer, then there exist a $x \in \mathbb{Z}_2^{n-1}$ such that $T(x) = -Id$ which is not possible since $\pi_1(M)$ is torsion-free.
Using the techniques introduced in Chapter 4 one can verify that the first Betti number of an orientable flat GHW manifold must be 0. □

In an analogous way, one can consider the class of non-orientable flat GHW manifolds M. For these manifolds, it is easy to prove that the first Betti number of M must be 0 or 1. In the latter case again the image of the holonomy representation is completely determined. In the former case there are more possibilities. In [51] the authors show that there are $n/2$ possibilities for n even and $(n + 1)/2$ possibilities for n odd. For more details we refer to [46] and [51].

To finish this section, we already note the following for flat GHW manifolds.

Theorem 6.3. *Let M be a flat GHW manifold and let $f : M \to M$ be a self-homotopy equivalence of M, then $N(f) = L(f)$.*

Proof: In [51], J.P.Rossetti and A. Szczepański proved that the outer automorphism group $\text{Out}(\pi_1(M))$ of the fundamental group of a flat GHW manifold is finite. Therefore it follows from [42] that each self-homotopy equivalence is homotopically periodic. And because of [34], we obtain that $N(f) = L(f)$.   □

6.2 Orientable flat GHW manifolds

The goal of this section is to prove the Anosov theorem for orientable flat GHW manifolds M, i.e. to show that the relation $N(f) = |L(f)|$ holds for any continuous map $f : M \to M$ (Theorem 6.10).

Let M be an orientable n-dimensional flat GHW manifold with fundamental group $E = \pi_1(M)$ and associated holonomy representation $T : F \to \text{Gl}(n, \mathbb{Z})$. It's image $T(F)$ is exactly the set of the 2^{n-1} diagonal matrices with 1 or -1 on the diagonal and such that the number of -1's is even (and the number of 1's is odd, as the dimension of M must be odd, Corollary 6.2).

The group $T(F)$ is hence generated by the set of diagonal matrices A_i $(1 \leq i \leq n - 1)$, where all the diagonal entries are -1, except the entry on the i-th row and column, which is 1. As a consequence, we can assume that the group E is generated by

$$(z_1, I_n), \ldots, (z_n, I_n), (a_1, A_1), \ldots (a_{n-1}, A_{n-1}) \qquad (6.1)$$

where $z_i \in \mathbb{Z}^n$ $(1 \leq i \leq n)$ and $a_i \in \mathbb{R}^n$ are appropriate translational parts. Let us, for the rest of this chapter, denote the k-th component of an element $b \in \mathbb{R}^n$ by b^k.

R. Miatello and J.P. Rossetti showed in [47, Lemma 1.4] that we can assume that the elements a_i^k of a_i are 0 or $\frac{1}{2}$. Although this result does not allow us to specify the translational parts a_i completely, we do already know that a_i^i must be $\frac{1}{2}$ for each i. Indeed, if $a_i^i = 0$, a simple computation shows that $(a_i, A_i) \cdot (a_i, A_i) = (a_i + A_i a_i, A_i^2) =$

$$\left(\begin{pmatrix} 0 \\ \vdots \\ 0 \\ \vdots \\ 0 \end{pmatrix}, I_n\right),$$ which would imply that E has torsion.

We will refer to a generating set (6.1) of E, with the $a_i^k = 0$ or $\frac{1}{2}$ as a suitable generating set.

Remark 6.4. *Let* $A_n = A_1 \cdot A_2 \cdot \ldots \cdot A_{n-1}$. *Then* A_n *is also a diagonal matrix with all diagonal entries equal to* -1, *except the last one which is 1 (since n is odd). Because of [47, Lemma 1.4] we can assume that there exists* $a_n \in \mathbb{R}^n$ *with components 0 or* $\frac{1}{2}$ *such that* $(a_n, A_n) \in E$.

As already mentioned above, in order to prove the Anosov theorem for flat orientable GHW manifolds, it is sufficient to deal with the continuous maps of M which are induced by a suitable affine endomorphism of $\mathbb{R}^n$. Therefore we need a full description of all affine endomorphisms $g : \mathbb{R}^n \to \mathbb{R}^n$ which are obtained in Corollary 1.11.

Lemma 6.5. *Let* M *be an orientable* n–*dimensional flat GHW manifold* $(n \geq 3)$ *and let* $(z_1, I_1), \ldots, (z_n, I_n), (a_1, A_1), \ldots, (a_{n-1}, A_{n-1})$ *be a suitable generating set of* $\pi_1(M) = E$. *Assume* θ *is a homomorphism of* E *and* $(\delta, \mathfrak{D}) \in \mathbb{R}^n \rtimes M_n(\mathbb{R})$ *is a suitable affine endomorphism (i.e.* $\forall \alpha \in E : \theta(\alpha) \cdot (\delta, \mathfrak{D}) = (\delta, \mathfrak{D}) \cdot \alpha)$. *Denoting the* (i,j)-*th entry of* $\mathfrak{D}$ *by* d_{ij} *we then have the following:*

1. *If there exists a* $j \in \{1, 2, \ldots, n\}$ *such that* $\theta(a_j, A_j) = (z, I_n)$, *with* $z \in \mathbb{Z}^n$ *(the image of some* (a_j, A_j) *is a pure translation), then for all* i, $(1 \leq i \leq n)$, $d_{ik} = 0$ *if* $k \neq j$ $(1 \leq k \leq n)$, *while* d_{ij} *is an even integer.*

2. *If there exists a* $j \in \{1, 2, \ldots, n\}$ *such that* $\theta(a_j, A_j) = (z, I_n)(b, B)$, *with* $z \in \mathbb{Z}^n$, $b \in \mathbb{R}^n$ *a translation consisting of 0's and* $\frac{1}{2}$'s *and* $B \neq I_n$ *a finite product of* A_i's *(the image of some* (a_j, A_j) *is not a pure translation), then there exists a* i $(1 \leq i \leq n)$ *such that* d_{ij} *is an odd integer and* $d_{ik} = 0$ *for all* $k \neq j$ $(1 \leq k \leq n)$.

Proof:

1. Since $\theta(a_j, A_j) \cdot (\delta, \mathfrak{D}) = (z, I_n) \cdot (\delta, \mathfrak{D}) = (z + \delta, \mathfrak{D})$ and $(\delta, \mathfrak{D}) \cdot (a_j, A_j) = (\delta + \mathfrak{D}a_j, \mathfrak{D}A_j)$, it follows that $\mathfrak{D} = \mathfrak{D}A_j$ and $z + \delta = \delta + \mathfrak{D}a_j$.

   Now, A_j only has $+1$ in the j-th column while the other diagonal entries are -1, this forces all the columns of $\mathfrak{D}$ to be zero, except the j-th column.

   For the translational parts we must have that $z + \delta = \delta + \mathfrak{D}a_j$. Since $a_j^j = \frac{1}{2}$ and only the j-th column of $\mathfrak{D}$ is non-zero, it follows that $(z^1 + d^1, \ldots, z^n + d^n)^t = (d^1 + \frac{1}{2}d_{1j}, \ldots, d^n + \frac{1}{2}d_{nj})^t$. So d_{ij} must be an even integer for all i ($1 \leq i \leq n$).

2. B is a finite product of A_i's, so B is also a diagonal matrix with 1's and -1's as diagonal entries. As above we denote the (i, j)-th entry of B by b_{ij}. Note that $(b, B) \cdot (b, B) = (b + Bb, I_n)$, from which it follows that $b + Bb$ can not be equal to zero. This implies that there exists a i ($1 \leq i \leq n$) such that $b_{ii} = 1$ and $b^i = \frac{1}{2}$.

   Since $\theta(a_j, A_j) \cdot (\delta, \mathfrak{D}) = (z + b, B) \cdot (\delta, \mathfrak{D}) = (z + b + B\delta, B\mathfrak{D})$ and $(\delta, \mathfrak{D}) \cdot (a_j, A_j) = (\delta + \mathfrak{D}a_j, \mathfrak{D}A_j)$, we have that $B\mathfrak{D} = \mathfrak{D}A_j$ and $z + b + B\delta = \delta + \mathfrak{D}a_j$.

   Since $b_{ii} = 1$, the i-th row of $B\mathfrak{D}$ equals the i-th row of $\mathfrak{D}$. Similarly the j-th column of $\mathfrak{D}A_j$ equals the j-th column of $\mathfrak{D}$ while the other columns of $\mathfrak{D}A_j$ are equal to minus the corresponding column of $\mathfrak{D}$. It follows that $d_{ik} = 0$ for all $k \neq j$ ($1 \leq k \leq n$).

   To show that d_{ij} is an odd integer, we look at the translational parts for which we know that $z + b + B\delta = \delta + \mathfrak{D}a_j$. Again, using $b_{ii} = 1$, the i-th component of the above equality reduces to

   $$z^i + b^i + d^i = d^i + \sum_{k=1}^{n} d_{ik}a_j^k = d^i + d_{ij}a_j^j = d^i + \frac{1}{2}d_{ij}.$$

   Since $b^i = \frac{1}{2}$ this shows that d_{ij} must be an odd integer.

   $\square$

Using the lemma above, we can now prove the following proposition in which we describe the linear parts of all suitable affine endomorphisms of $\mathbb{R}^n$.

Proposition 6.6. *Let M be an orientable n-dimensional flat GHW manifold with fundamental group $\pi_1(M) = E$ ($n \geq 3$). Let θ be a homomorphism of E and $(\delta, \mathfrak{D}) \in \mathbb{R}^n \rtimes M_n(\mathbb{R})$ be a suitable affine endomorphism. Then*

1. $\mathfrak{D}$ *is either the zero* $n \times n$ *matrix* 0_n *or*

2. $\mathfrak{D}$ *is an element of* $\mathrm{Gl}(n, \mathbb{Q})$ *such that in each row and each column of* $\mathfrak{D}$ *there is exactly one non-zero element, which is an odd integer.*

Proof: To prove this proposition we distinguish three cases depending on the number of (a_i, A_i)'s which are mapped onto a pure translation.

Case 1: Suppose that θ is a homomorphism of E such that two or more of the images of $(a_1, A_1), \ldots, (a_n, A_n)$ are pure translations. So there exist i and j, $i \neq j$, for which $\theta(a_i, A_i)$ and $\theta(a_j, A_j)$ are pure translations. Then Lemma 6.5 implies that $\mathfrak{D} = 0_n$.

Case 2: Suppose that θ is a homomorphism of E such that just one of the images of $(a_1, A_1), \ldots, (a_n, A_n)$ is a pure translation. So part one of Lemma 6.5 implies that all the elements of $\mathfrak{D}$ are even integers. But there also has to exist a j such that $\theta(a_j, A_j)$ is not a pure translation. So part two of Lemma 6.5 implies that there is a i such that d_{ij} is an odd integer. Since this gives a contradiction, we conclude that no such homomorphism exists.

Case 3: Suppose that θ is a homomorphism of E such that none of the images of $(a_1, A_1), \ldots (a_n, A_n)$ is a pure translation. In this situation, we can determine $\mathfrak{D}$ completely. Namely, since $\theta(a_1, A_1)$ is not a pure translation, Lemma 6.5 implies that there exists a i_1 such that $d_{i_1 1}$ is odd, while the other elements of the i_1-th row are zero. Doing the same for (a_2, A_2) we obtain a i_2 such that $d_{i_2 2}$ is odd, while the other elements of the i_2-th row are zero. Clearly $i_2 \neq i_1$, otherwise we would have that $d_{i_2 2}$ is zero on the one hand and an odd integer on the other hand. This can be done for all the images of $(a_1, A_1), \ldots (a_{n-1}, A_{n-1})$ and (a_n, A_n), so we have completely determined $\mathfrak{D}$ as a matrix with exactly one non-zero, odd entry in each row and column.

$\square$

Remark 6.7. *In the above proposition, we did not mention the image of the pure translations. But since the $\mathfrak{D}$ always consist of integers, one can easily show that the image of a pure translation under the homomorphism again must be a pure translation.*

Now that we have a clear view on the different possibilities for the linear parts of the suitable affine endomorphisms $(\delta, \mathfrak{D})$, we can start to use Theorem 2.5.

As a first step, in the following lemma we calculate the determinants which appear in Theorem 2.5 and in a second lemma we determine the signs of these determinants. To a $(n \times n)$-matrix $\mathfrak{D}$ with in each row and each column exactly one non-zero element (as in the second part of Proposition 6.6), one can associate an unique permutation μ of n elements. Namely, for any $i = 1, 2, \ldots, n$ let $\mu(i)$ be the unique index such that $d_{i\mu(i)} \neq 0$. Clearly μ is a permutation of n elements and μ has a unique cycle decomposition.

Lemma 6.8. *Suppose B is any diagonal matrix whose diagonal entries b_{ii} are 1's and -1's.*

1. *If $\mathfrak{D} = 0_n$ is the zero $(n \times n)$-matrix, then $\det(I_n - B\mathfrak{D}) = 1$ for all possible B.*
2. *Let $\mathfrak{D}$ be a $(n \times n)$-matrix, such that in each row and each column of $\mathfrak{D}$ there is exactly one non-zero element and let μ be the associated permutation. Let the cycle decomposition of μ be*

$$(l_1^1 \ l_2^1 \ \cdots \ l_{p_1}^1)(l_1^2 \ l_2^2 \ \cdots \ l_{p_2}^2) \cdots (l_1^r \ l_2^r \ \cdots \ l_{p_r}^r).$$

Then we have that

$$\det(I_n - B\mathfrak{D})$$
$$= \det(B) \times \prod_{i=1}^{r} (b_{l_1^i l_1^i} b_{l_2^i l_2^i} \cdots b_{l_{p_i}^i l_{p_i}^i} - d_{l_1^i \mu(l_1^i)} d_{l_2^i \mu(l_2^i)} \cdots d_{l_{p_i}^i \mu(l_{p_i}^i)})$$
$$= \det(B) \times \prod_{i=1}^{r} (b_{l_1^i l_1^i} b_{l_2^i l_2^i} \cdots b_{l_{p_i}^i l_{p_i}^i} - d_{l_1^i l_2^i} d_{l_2^i l_3^i} \cdots d_{l_{p_i}^i l_1^i}).$$

Proof:

1. Trivial.
2. Since $B^2 = I_n$, we have that $\det(I_n - B\mathfrak{D}) = \det(B) . \det(B - \mathfrak{D})$. Now we show that

$$\det(B - \mathfrak{D}) = \prod_{i=1}^{r} (b_{l_1^i l_1^i} b_{l_2^i l_2^i} \cdots b_{l_{p_i}^i l_{p_i}^i} - d_{l_1^i \mu(l_1^i)} d_{l_2^i \mu(l_2^i)} \cdots d_{l_{p_i}^i \mu(l_{p_i}^i)}).$$

We do this by induction on n, the case $n = 1$ being trivial. Suppose the formula holds for $(n-1) \times (n-1)$-matrices $(n \geq 2)$, then we distinguish two cases:

a) $d_{nn} \neq 0$ or equivalently $\mu(n) = n$. Then $\det(B - \mathfrak{D}) = (b_{nn} - d_{n\mu(n)}) \det(B' - \mathfrak{D}')$, where B' (resp. $\mathfrak{D}'$) is obtained from the matrix B (resp. $\mathfrak{D}$) by deleting the last row and column. By applying the induction hypothesis to $B' - \mathfrak{D}'$ we obtain the result.

b) $d_{nn} = 0$. Then there exists a unique k such that $d_{nk} \neq 0$ (or $\mu(n) = k$) and a unique l with $d_{ln} \neq 0$ (or $\mu(l) = n$). So in the cycle decomposition of μ, there is a cycle of the form $(l_1 \cdots l\ n\ k \cdots l_p)$.

In the last column of the matrix $B - \mathfrak{D}$ there are two non-zero elements, namely b_{nn} and $-d_{ln}$.

Since $b_{nn} = \pm 1$ we can create a zero in the last column of the l-th row by adding to the l-th row $b_{nn} \cdot d_{ln}$ times the last row. In the computation below, this operation is used in the step indicated with $(*)$.

In this calculation, we again use B' and $\mathfrak{D}'$ to denote the matrices obtained by deleting the last row and column of B and $\mathfrak{D}$ respectively. Note that the l-th row of $\mathfrak{D}'$ and the k-th column of $\mathfrak{D}'$ only consists of zeros and that in each other row and each other column of $\mathfrak{D}'$ there is exactly one non-zero element.

Also, we need $\mathfrak{D}''$ obtained from $\mathfrak{D}'$ by changing the k-th component of the l-th row of $\mathfrak{D}'$ to $-b_{nn}d_{ln}d_{nk}$. Since $\mathfrak{D}'$ is from the above form, we then have that in each row and each column of $\mathfrak{D}''$ there is exactly one non-zero element. So

$$\det(B - \mathfrak{D}) = \det \begin{pmatrix} & & & 0 \\ & & & \vdots \\ & & & 0 \\ & B' - \mathfrak{D}' & & -d_{ln} \\ & & & 0 \\ & & & \vdots \\ & & & 0 \\ 0 \cdots 0 & -d_{nk}\ 0 & \cdots 0 & b_{nn} \end{pmatrix}$$

$$\stackrel{(*)}{=} \det \begin{pmatrix} & & & 0 \\ & B' - \mathfrak{D}'' & & \vdots \\ & & & 0 \\ 0 \cdots 0 & -d_{nk}\ 0 & \cdots 0 & b_{nn} \end{pmatrix}$$

$$= b_{nn} \det(B' - \mathfrak{D}'') \tag{6.2}$$

Now we can associate a permutation μ'' of $n-1$ elements to $\mathfrak{D}''$. The cycle decomposition of μ'' is obtained from the cycle decomposition of μ by replacing in this decomposition the cycle $(l_1 \cdots l \, n \, k \cdots l_p)$ which contains n, with the cycle $(l_1 \cdots l \, k \cdots l_p)$.

The induction hypothesis now applies to the matrix $B' - \mathfrak{D}''$ and it follows that all factors in the expansion of $\det(B' - \mathfrak{D}'')$ except the one containing the terms of the l-th row are of the desired form. The exceptional factor containing the elements of the l-th row and corresponding to the cycle $(l_1 \cdots l \, k \cdots l_p)$ is of the following form:

$$b_{l_1 l_1} \cdots b_{ll} \cdots b_{l_p l_p} - d_{l_1 \mu(l_1)} \cdots (b_{nn} d_{ln} d_{nk}) \cdots d_{l_p \mu(l_p)}$$

We can multiply the factor above with b_{nn} to get

$$b_{nn} b_{l_1 l_1} \cdots b_{ll} \cdots b_{l_p l_p} - b_{nn} d_{l_1 \mu(l_1)} \cdots (b_{nn} d_{ln} d_{nk}) \cdots d_{l_p \mu(l_p)}$$

which equals $(n = \mu(l)$ and $\mu(n) = k)$

$$b_{l_1 l_1} \cdots b_{ll} b_{nn} \cdots b_{l_p l_p} - d_{l_1 \mu(l_1)} \cdots d_{l\mu(l)} d_{n\mu(n)} \cdots d_{l_p \mu(l_p)}$$

This part of the expansion of (6.2) together with the other factors show that $\det(B - \mathfrak{D})$ is of the desired form.

$\square$

Note that the number of factors of the determinants in the previous lemma does not depend on B, but only on $\mathfrak{D}$ (in fact, only on the form of D determined by μ). So for a given $\mathfrak{D}$, with $\mathfrak{D}$ as in the second case of Lemma 6.5, and any diagonal matrix B consisting of 1's and -1's, we obtain that $\det(I_n - B\mathfrak{D}) = \det(B)(\pm 1 - x_1) \cdots (\pm 1 - x_k)$. Here $x_1, \ldots, x_k \in 1 + 2\mathbb{Z}$ and the ± 1's depend on the choice of B. In this perspective the following lemma is crucial.

Lemma 6.9. *Fix an integer $k \geq 1$ and $x_1, \ldots, x_k \in 1 + 2\mathbb{Z}$. Then*

$$\text{either } (\epsilon_1 - x_1) \cdots (\epsilon_k - x_k) \geq 0 \text{ for all possible } \epsilon_1, \ldots, \epsilon_k \in \{-1, 1\}$$
$$\text{or } (\epsilon_1 - x_1) \cdots (\epsilon_k - x_k) \leq 0 \text{ for all possible } \epsilon_1, \ldots, \epsilon_k \in \{-1, 1\}.$$

<u>Proof:</u> Suppose there exists $\epsilon_1, \ldots, \epsilon_k$ and $\epsilon_1', \ldots, \epsilon_k'$ such that $(\epsilon_1 - x_1) \cdots (\epsilon_k - x_k) > 0$ and $(\epsilon_1' - x_1) \cdots (\epsilon_k' - x_k) < 0$. This is only possible if there exist a j for which $1 - x_j > 0$ and $-1 - x_j < 0$ or conversely

$1 - x_j < 0$ and $-1 - x_j > 0$. But in the former case we quickly see that $x_j = 0$ which is not possible and in the latter case we obtain the contradiction that $x_j > 1$ and $x_j < -1$.   □

We are now ready to prove the Anosov theorem for flat orientable GHW manifolds.

Theorem 6.10. *Let $n \geq 3$ be an odd integer and M a (flat) orientable n-dimensional generalized Hantzsche-Wendt manifold. Then for each continuous map $f : M \to M$ we have that $N(f) = |L(f)|$.*

Proof: Suppose $f : M \to M$ is a continuous map on M. Due to Corollary 1.11 we know that f is homotopic to a map g induced by an suitable affine endomorphism $(\delta, \mathfrak{D})$ of $\mathbb{R}^n$ and due to Proposition 6.6 we know how $\mathfrak{D}$ looks like. Since the Nielsen and Lefschetz number are homotopy invariants it suffices to prove the theorem for the map g. We use Theorem 2.5 to verify that $N(g) = |L(g)|$. Therefore we have to calculate $\det(I_n - T(x)\mathfrak{D})$ for each $x \in F$. Note that for each $x \in F$, $T(x)$ is a diagonal matrix whose diagonal entries consist of an even numbers of -1's while the others are 1 and so $\det(T(x)) = 1$. Therefore we can apply Lemma 6.8 and Lemma 6.9 to the determinants $\det(I_n - T(x)\mathfrak{D})$ which finishes the proof of the theorem.   □

Part III

The Anosov theorem in small dimensions

In this part we verify what we know about the Anosov theorem for infra-nilmanifolds in small dimensions, i.e. up to dimension 4. In this way we show that many infra-nilmanifolds are already covered by the original result of Anosov and our results of the previous part. However there are still examples of infra-nilmanifolds which are not yet covered by the known theorems. By examining these manifolds we hope to find some tracks for further research, in order to close the remaining lack of theoretic results.

In a first chapter we examine the flat manifolds and we use [7] for a complete description of these flat manifolds. For flat manifolds we have more results which we can apply and the calculations are a lot easier. This results in a complete picture for dimension 3, but in dimension 4 only 55 of the 74 flat manifolds are covered. The 19 remaining manifolds can be divided in two classes. To be specific 4-dimensional flat manifolds with holonomy group $\mathbb{Z}_2 \oplus \mathbb{Z}_2$ and 4-dimensional flat manifolds with non-abelian holonomy group. In the last two sections we observe some intriguing results for these two classes.

In a second chapter we focus on infra-nilmanifolds up to dimension 4. We use [12] for a complete description of these infra-nilmanifolds. We also include the infra-nilmanifold introduced in [18] which is missing in the previous description. Firstly, the calculations for these manifolds are much more complicated than for flat manifolds. Therefore we start this chapter by formally showing how the calculations need to be done. Secondly we have less theoretic results which we can apply (for instance Proposition 5.12) and so lot of work had to be done (or many possibilities for further research are left over).

For all the manifolds which are not covered by the theoretic results, we present in this part a counterexample (in the case the Anosov theorem does not hold for the specific manifold) or a proof (in the opposite case). We demonstrate the calculations that need to be done throughout several well chosen examples. All this results in a complete answer to the question for which infra-nilmanifolds up to dimension 4 the Anosov theorem holds.

Chapter 7

Flat manifolds

In the sequel we need to refer explicitly to specific flat manifolds. As any flat manifold is determined by its fundamental group, this boils down to finding a way of explicitly referring to the associated Bieberbach group. We have chosen to use the numbers which can be found on "the page of low dimensional Bieberbach groups" at the CARAT website [7]. This allows us to say that the orientable GHW manifold corresponds to the number 10.1.1. Actually, this number refers to the $\mathbb{Z}$–class to which the fundamental group of the orientable GHW manifold belongs. Two Bieberbach groups can belong to the same $\mathbb{Z}$–class, by which one means that the corresponding integral holonomy representations are conjugated over $\mathbb{Z}$. If we need to talk about different Bieberbach groups in the same $\mathbb{Z}$–class, we will use the same order as on the CARAT website. E.g. when one asks the CARAT system for the Bieberbach groups of dimension 3 with holonomy group $\mathbb{Z}_2 \times \mathbb{Z}_2$, one receives as an answer:

```
#g2 % 1-th torsion free group in Z-class min.10.1.1
4    /2       % generator
  -2 0   0 0
   0 2   0 1
   0 0  -2 1
   0 0   0 2
4    /2       % generator
   2  0   0 1
   0 -2   0 0
   0  0  -2 0
   0  0   0 2
2^2  % order of the group
```

```
#g2 % 1-th torsion free group in Z-class min.9.1.1
4    /2       % generator
 -2 0 0 0
  0 2 0 0
  0 0 2 1
  0 0 0 2
4    /2       % generator
  2  0 0 1
  0 -2 0 0
  0  0 2 0
  0  0 0 2
2^2  % order of the group
#g2 % 2-th torsion free group in Z-class min.9.1.1
4    /2       % generator
 -2 0 0 0
  0 2 0 1
  0 0 2 1
  0 0 0 2
4    /2       % generator
  2  0 0 1
  0 -2 0 0
  0  0 2 0
  0  0 0 2
2^2  % order of the group
```

This table contains for each of the Bieberbach groups a set of generators, which should in any case be completed with a set of generators for the integral lattice $\mathbb{Z}^n$ (here $\mathbb{Z}^3$). The table above shows that the orientable GHW belongs to the $\mathbb{Z}$-class with number 10.1.1. while the two non-orientable ones both belong to the $\mathbb{Z}$–class 9.1.1.

Finally, we note here that in literature one can find other classifications of the Bieberbach groups and so implicitly also of the flat manifolds, see for instance [3]. In this book, Bieberbach groups are divided into families, families into crystal systems, ... and also finally into $\mathbb{Z}$–classes (before the really final step, up to isomorphism, of course).

Where useful, we will always mention these $\mathbb{Z}$-class numbers (and the exact number inside the $\mathbb{Z}$-class when needed). In this way, one can use [7] to find a precise description of the fundamental group of the respective manifold. We prefer to base our referring system on [7], instead of e.g. [3], because of the overall availability of [7].

7.1 General overview in dimension 3 and 4

For flat manifolds up to dimension 2, Anosov provided a complete picture, since we only have the circle in dimension 1 and the torus and the Klein bottle in dimension 2. By the theorems proved in the previous part of this book, we also get a complete overview of the situation in dimension 3, where there are 10 flat manifolds. We summarize in the following table.

Dimension 3				
Holonomy group	# mani- folds	Orientable		Non orientable
		An th. holds	An.th. holds not	An.th. holds not
1	1	1	0	0
$\mathbb{Z}_2$	3	0	1	2
$\mathbb{Z}_3$	1	1	0	0
$\mathbb{Z}_4$	1	1	0	0
$\mathbb{Z}_6$	1	1	0	0
$\mathbb{Z}_2 \oplus \mathbb{Z}_2$	3	1	0	2
Total	10	5	1	4

We explain which information can be found in this table (and the corresponding one for flat manifolds in dimension 4 on page 98).

In the first column we list all possible holonomy groups and in the second column we give the number of corresponding manifolds. In the following columns we then summarize what we know, about the validity of the Anosov theorem for the specific manifolds. We do this by giving respectively the number of orientable manifolds for which the Anosov theorem holds; the number of orientable manifolds for which the Anosov theorem does not hold and the number of non-orientable manifolds (for which the Anosov theorem never holds because of Proposition 3.7).

We conclude that in dimension 3 there are 5 flat manifolds for which the Anosov theorem holds: the torus which is covered by Theorem 2.3, the 3 manifolds with cyclic holonomy group for which Theorem 5.4 holds and the orientable GHW manifold for which Theorem 6.10 holds.

From the 5 manifolds for which the Anosov theorem does not hold, there are 4 non-orientable manifolds for which Proposition 3.7 holds and one orientable manifold with holonomy group $\mathbb{Z}_2$ (7.1.1) for which Proposition 5.12 holds.

In contrary to dimension 3, in dimension 4 there are some manifolds which are not covered by the theorems of the previous chapters. From the 74 flat manifolds in dimension 4 there are 55 manifolds covered by the known results which mainly comes from Proposition 3.7 concerning non-orientable flat manifolds. So 19 manifolds still need to be examined by 'hand'. The results are summarized in the table below.

Dimension 4				
Holonomy group	# mani- folds	Orientable		Non orientable
		An th. holds	An.th. holds not	An.th. holds not
1	1	1	0	0
$\mathbb{Z}_2$	5	0	2	3
$\mathbb{Z}_3$	2	2	0	0
$\mathbb{Z}_4$	6	2	0	4
$\mathbb{Z}_6$	4	1	0	3
$\mathbb{Z}_2 \oplus \mathbb{Z}_2 \oplus \mathbb{Z}_2$	12	0	0	12
$\mathbb{Z}_2 \oplus \mathbb{Z}_2$	26	4	5	17
$\mathbb{Z}_2 \oplus \mathbb{Z}_2 \oplus \mathbb{Z}_3$	1	0	0	1
$\mathbb{Z}_2 \oplus \mathbb{Z}_4$	3	0	0	3
D_8	7	2	2	3
S_3	3	0	3	0
$\mathbb{Z}_2 \oplus S_3$	1	0	1	0
A_4	2	2	0	0
$\mathbb{Z}_2 \oplus A_4$	1	0	0	1
Total	74	14	13	47

Note that the first half of the table contains these manifolds for which we have a complete picture by the theorems of the previous part. Namely the torus, the 17 manifolds with cyclic holonomy group which are covered by Proposition 3.7, Theorem 5.4 and Proposition 5.12 and the 12 non-orientable GHW manifolds which are also covered by Proposition 3.7.

In the second half of the table we find 25 non-orientable manifolds for which we again can apply Proposition 3.7 to obtain that the Anosov theorem does not hold.

The remaining 19 manifolds which need a more detailed examination can be split up into two classes: 9 flat manifolds with $\mathbb{Z}_2 \oplus \mathbb{Z}_2$ as holonomy group and 10 flat manifolds with a non-abelian holonomy group. Both of these classes are very interesting. The first class, for instance is exactly in between the class of flat manifolds with $\mathbb{Z}_2$ as holonomy group and the class of generalized Hantzsche-Wendt manifolds, two classes for which we already have a complete picture by Theorem 5.4, Proposition 5.12 and Theorem 6.10. Secondly in the previous chapters we mainly focused on infra-nilmanifolds with an abelian holonomy group (see Chapter 5 and 6, which explains why the second class could be interesting). In the following two sections we discuss the two classes in detail.

7.2 Flat manifolds in dimension 4 with $\mathbb{Z}_2 \oplus \mathbb{Z}_2$ as holonomy group

Let M be an orientable flat manifold in dimension 4 with $\mathbb{Z}_2 \oplus \mathbb{Z}_2$ as holonomy group. As already stressed before we only have to consider the orientable manifolds. Then we may assume that the fundamental group $\pi_1(M)$ is generated by $(e_1, I_4), (e_2, I_4), (e_3, I_3), (e_4, I_4), (a_1, A_1)$ and (a_2, A_2), with 1 in the i-th place for e_i and 0 everywhere else; $A_1, A_2 \in \mathrm{Gl}(4, \mathbb{Z})$ commuting matrices of order 2 and a_1, a_2 suitable translational parts. Since A_1 and A_2 commute we know that there exist a $P \in \mathrm{Gl}(n, \mathbb{Q})$ such that

$$PA_1P^{-1} = B_1 = \begin{pmatrix} 1 & 0 & 0 & 0 \\ 0 & 1 & 0 & 0 \\ 0 & 0 & -1 & 0 \\ 0 & 0 & 0 & -1 \end{pmatrix} \text{ and } PA_2P^{-1} = B_2 = \begin{pmatrix} 1 & 0 & 0 & 0 \\ 0 & -1 & 0 & 0 \\ 0 & 0 & -1 & 0 \\ 0 & 0 & 0 & 1 \end{pmatrix}.$$

The multiplicity of -1 must indeed be 2 since M needs to be orientable (see also Proposition 1.6).

As explained before, in order to verify the validity of the Anosov theorem for a specific manifold, we need a description of all suitable affine endomorphisms $(\delta, \mathfrak{D}) \in \mathbb{R}^4 \rtimes \mathrm{Endo}(\mathbb{R}^4)$. To construct all these endomorphisms we actually have to consider all the possible homomorphisms $\theta : \pi_1(M) \to \pi_1(M)$, since for each $(a, A) \in \pi_1(M)$:

$$\theta((a, A))(\delta, \mathfrak{D}) = (\delta, \mathfrak{D})(a, A)$$

Suppose that $\theta((a, A)) = (b, B)$, then this boils down to

$$(b + B\delta, B\mathfrak{D}) = (\delta + \mathfrak{D}a, \mathfrak{D}A)$$

This equation gives us two types of conditions which need to be satisfied, namely one type concerning the linear parts and another concerning the translational parts. Therefore in a first step we construct all possible matrices $\mathfrak{D}$ such that $B\mathfrak{D} = \mathfrak{D}A$. In a second step we then look at the consequences of the fact that $\delta + \mathfrak{D}a = b + B\delta$. This second step has to be done separately for each of the 9 manifolds under study, but the work of the first step can be simplified a lot because of the existence of P. Indeed we only have to search for all matrices $\mathfrak{D}$ such that $PBP^{-1}\mathfrak{D} = \mathfrak{D}PAP^{-1}$ since

$$PBP^{-1}\mathfrak{D} = \mathfrak{D}PAP^{-1} \Leftrightarrow BP^{-1}\mathfrak{D}P = P^{-1}\mathfrak{D}PA$$

So $P^{-1}\mathfrak{D}P$ gives us then a full description of all possible matrices. The list below contains 16 possible matrices, since in order to construct all possible homomorphisms θ, we consider the 16 possible conditions $X\mathfrak{D} = \mathfrak{D}B_i$, with $i = 1, 2$ and $X \in \{I_4, B_1, B_2, B_1B_2\}$.

We could (and perhaps should) in fact further partition each of these sixteen cases into 4 more cases, depending upon the four possibilities $X\mathfrak{D} = \mathfrak{D}I_4$, for $X = I_1$, B_1, B_2 or B_1B_1. This corresponds to the fact that the image of a translation under the homomorphism θ does not need to be a translation. These extra conditions will of course not lead to 'new possibilities' for $\mathfrak{D}$, but they will only imply that extra zeroes will appear in the possible $\mathfrak{D}$'s.

The 16 matrices are the following. Note that the d_{ij} will be specified later on.

$\begin{aligned}\mathfrak{D}_1 &= \mathfrak{D}_1 B_1 \\ \mathfrak{D}_1 &= \mathfrak{D}_1 B_2\end{aligned}$	$\begin{aligned}\mathfrak{D}_2 &= \mathfrak{D}_2 B_1 \\ B_1\mathfrak{D}_2 &= \mathfrak{D}_2 B_2\end{aligned}$
$\mathfrak{D}_1 = \begin{pmatrix} d_{11} & 0 & 0 & 0 \\ d_{21} & 0 & 0 & 0 \\ d_{31} & 0 & 0 & 0 \\ d_{41} & 0 & 0 & 0 \end{pmatrix}$	$\mathfrak{D}_2 = \begin{pmatrix} d_{11} & 0 & 0 & 0 \\ d_{21} & 0 & 0 & 0 \\ 0 & d_{32} & 0 & 0 \\ 0 & d_{42} & 0 & 0 \end{pmatrix}$
$\begin{aligned}\mathfrak{D}_3 &= \mathfrak{D}_3 B_1 \\ B_2\mathfrak{D}_3 &= \mathfrak{D}_3 B_2\end{aligned}$	$\begin{aligned}\mathfrak{D}_4 &= \mathfrak{D}_4 B_1 \\ B_1 B_2\mathfrak{D}_4 &= \mathfrak{D}_4 B_2\end{aligned}$
$\mathfrak{D}_3 = \begin{pmatrix} d_{11} & 0 & 0 & 0 \\ 0 & d_{22} & 0 & 0 \\ 0 & d_{32} & 0 & 0 \\ d_{41} & 0 & 0 & 0 \end{pmatrix}$	$\mathfrak{D}_4 = \begin{pmatrix} d_{11} & 0 & 0 & 0 \\ 0 & d_{22} & 0 & 0 \\ d_{31} & 0 & 0 & 0 \\ 0 & d_{42} & 0 & 0 \end{pmatrix}$
$\begin{aligned}B_1\mathfrak{D}_5 &= \mathfrak{D}_5 B_1 \\ \mathfrak{D}_5 &= \mathfrak{D}_5 B_2\end{aligned}$	$\begin{aligned}B_1\mathfrak{D}_6 &= \mathfrak{D}_6 B_1 \\ B_1\mathfrak{D}_6 &= \mathfrak{D}_6 B_2\end{aligned}$
$\mathfrak{D}_5 = \begin{pmatrix} d_{11} & 0 & 0 & 0 \\ d_{21} & 0 & 0 & 0 \\ 0 & 0 & 0 & d_{34} \\ 0 & 0 & 0 & d_{44} \end{pmatrix}$	$\mathfrak{D}_6 = \begin{pmatrix} d_{11} & 0 & 0 & 0 \\ d_{21} & 0 & 0 & 0 \\ 0 & 0 & d_{33} & 0 \\ 0 & 0 & d_{43} & 0 \end{pmatrix}$
$\begin{aligned}B_1\mathfrak{D}_7 &= \mathfrak{D}_7 B_1 \\ B_2\mathfrak{D}_7 &= \mathfrak{D}_7 B_2\end{aligned}$	$\begin{aligned}B_1\mathfrak{D}_8 &= \mathfrak{D}_8 B_1 \\ B_1 B_2\mathfrak{D}_8 &= \mathfrak{D}_8 B_2\end{aligned}$
$\mathfrak{D}_7 = \begin{pmatrix} d_{11} & 0 & 0 & 0 \\ 0 & d_{22} & 0 & 0 \\ 0 & 0 & d_{33} & 0 \\ 0 & 0 & 0 & d_{44} \end{pmatrix}$	$\mathfrak{D}_8 = \begin{pmatrix} d_{11} & 0 & 0 & 0 \\ 0 & d_{22} & 0 & 0 \\ 0 & 0 & 0 & d_{34} \\ 0 & 0 & d_{43} & 0 \end{pmatrix}$

$B_2\mathfrak{D}_9 = \mathfrak{D}_9 B_1$ $\mathfrak{D}_9 = \mathfrak{D}_9 B_2$ $\mathfrak{D}_9 = \begin{pmatrix} d_{11} & 0 & 0 & 0 \\ 0 & 0 & 0 & d_{24} \\ 0 & 0 & 0 & d_{34} \\ d_{41} & 0 & 0 & 0 \end{pmatrix}$	$B_2\mathfrak{D}_{10} = \mathfrak{D}_{10} B_1$ $B_1\mathfrak{D}_{10} = \mathfrak{D}_{10} B_2$ $\mathfrak{D}_{10} = \begin{pmatrix} d_{11} & 0 & 0 & 0 \\ 0 & 0 & 0 & d_{24} \\ 0 & 0 & d_{33} & 0 \\ 0 & d_{42} & 0 & 0 \end{pmatrix}$
$B_2\mathfrak{D}_{11} = \mathfrak{D}_{11} B_1$ $B_2\mathfrak{D}_{11} = \mathfrak{D}_{11} B_2$ $\mathfrak{D}_{11} = \begin{pmatrix} d_{11} & 0 & 0 & 0 \\ 0 & 0 & d_{23} & 0 \\ 0 & 0 & d_{33} & 0 \\ d_{41} & 0 & 0 & 0 \end{pmatrix}$	$B_2\mathfrak{D}_{12} = \mathfrak{D}_{12} B_1$ $B_1 B_2\mathfrak{D}_{12} = \mathfrak{D}_{12} B_2$ $\mathfrak{D}_{12} = \begin{pmatrix} d_{11} & 0 & 0 & 0 \\ 0 & 0 & d_{23} & 0 \\ 0 & 0 & 0 & d_{34} \\ 0 & d_{42} & 0 & 0 \end{pmatrix}$
$B_1 B_2\mathfrak{D}_{13} = \mathfrak{D}_{13} B_1$ $\mathfrak{D}_{13} = \mathfrak{D}_{13} B_2$ $\mathfrak{D}_{13} = \begin{pmatrix} d_{11} & 0 & 0 & 0 \\ 0 & 0 & 0 & d_{24} \\ d_{31} & 0 & 0 & 0 \\ 0 & 0 & 0 & d_{44} \end{pmatrix}$	$B_1 B_2\mathfrak{D}_{14} = \mathfrak{D}_{14} B_1$ $B_1\mathfrak{D}_{14} = \mathfrak{D}_{14} B_2$ $\mathfrak{D}_{14} = \begin{pmatrix} d_{11} & 0 & 0 & 0 \\ 0 & 0 & 0 & d_{24} \\ 0 & d_{32} & 0 & 0 \\ 0 & 0 & d_{43} & 0 \end{pmatrix}$
$B_1 B_2\mathfrak{D}_{15} = \mathfrak{D}_{15} B_1$ $B_2\mathfrak{D}_{15} = \mathfrak{D}_{15} B_2$ $\mathfrak{D}_{15} = \begin{pmatrix} d_{11} & 0 & 0 & 0 \\ 0 & 0 & d_{23} & 0 \\ 0 & d_{32} & 0 & 0 \\ 0 & 0 & 0 & d_{44} \end{pmatrix}$	$B_1 B_2\mathfrak{D}_{16} = \mathfrak{D}_{16} B_1$ $B_1 B_2\mathfrak{D}_{16} = \mathfrak{D}_{16} B_2$ $\mathfrak{D}_{16} = \begin{pmatrix} d_{11} & 0 & 0 & 0 \\ 0 & 0 & d_{23} & 0 \\ d_{31} & 0 & 0 & 0 \\ 0 & 0 & d_{43} & 0 \end{pmatrix}$

Remark 7.1. *Note that this method and matrices can be adapted to cover any orientable flat manifold with holonomy group $\mathbb{Z}_2 \oplus \mathbb{Z}_2$. For such manifolds the $\mathbb{Q}$-irreducible components are one-dimensional but their multiplicity needs no longer to be 1 (as is the case for the manifolds above). Using the techniques described above one again obtains 16 possible linear parts but now the d_{ij} have to be replaced by matrices*

D_{ij} of the 'right' dimensions. This comes from the fact that there are $\mathbb{Q}$-irreducible components with multiplicity greater then one.

So there are 16 cases which we need to examine, but for 6 cases we can already state that the Anosov relation holds. Indeed, if we calculate the determinants in Theorem 2.5 for $\mathfrak{D}_1, \mathfrak{D}_2, \mathfrak{D}_9$ and $\mathfrak{D}_{16}$ we always obtain $1 - d_{11}$. So whatever the value of d_{11} is, we always have that the Anosov relation holds. Secondly for the cases $\mathfrak{D}_{12}$ and $\mathfrak{D}_{14}$ the conditions in Proposition 2.6 are satisfied which again implies that the Anosov relation holds. So, in the discussion below we no longer have to include these linear parts.

Evaluating the second type of conditions on the translational parts leads to some interesting results for the remaining linear parts. This comes from the fact that, as is for instance also done in Chapter 6, in this step we specify the entries of the matrices mentioned above. Moreover in some cases we have to exclude many of the matrices since the entries of the matrices can not satisfy the conditions coming from the translational parts. This is nicely demonstrated for this first manifold.

Let E_1 be the first Bieberbach group in 22.1.1 which is generated by

$$(e_i, I_4), \qquad (a_1, A_1) = (\begin{pmatrix} 0 \\ \frac{1}{2} \\ \frac{1}{2} \\ 0 \end{pmatrix}, \begin{pmatrix} 1 & 0 & 0 & 0 \\ 0 & 1 & 0 & 0 \\ 0 & 0 & -1 & 0 \\ 0 & 0 & 0 & -1 \end{pmatrix})$$

$$\text{and } (a_2, A_2) = (\begin{pmatrix} 0 \\ 0 \\ 0 \\ \frac{1}{2} \end{pmatrix}, \begin{pmatrix} 1 & 0 & 0 & 0 \\ 0 & -1 & 0 & 0 \\ 0 & 0 & -1 & 0 \\ 0 & 0 & 0 & 1 \end{pmatrix})$$

with 1 in the i-th place and 0 everywhere else for e_i; $i = 1, \ldots 4$. To be precise we slightly transformed the matrices found on [7] to avoid the explicit use of the matrix P. Then $M_1 = E_1 \backslash \mathbb{R}^4$ is indeed one of the manifolds under study.

Hereafter, we only focus on the linear parts $\mathfrak{D}$, and not on the translational parts δ, since the validity of the Anosov theorem only depends on $\mathfrak{D}$. For this manifold we have that from the 10 remaining possibilities only $\mathfrak{D}_7, \mathfrak{D}_8, \mathfrak{D}_{10}, \mathfrak{D}_{12}, \mathfrak{D}_{14}$ and $\mathfrak{D}_{15}$ are suitable linear parts under the extra condition that each non-zero entry is an odd integer except for d_{11} which can be any integer. Note that possibly the same phenomenon can happen for the 6 cases that we do not consider.

Let us demonstrate the calculations that need to be done for instance for $\mathfrak{D}_7$. In that case we actually assume that $\theta(a_i, A_i) = (z_i + a_i, A_i)$ with $z_i \in \mathbb{Z}_4$ and $i = 1, 2$. So from the second type of conditions on the translational parts, we obtain that $z_i + a_i + A_i\delta = \delta + \mathfrak{D}_7 a_i$ or equivalently that

$$(I_4 - A_i)\delta + (\mathfrak{D}_7 - I_4)a_i \in \mathbb{Z}^4.$$

With $\delta = (t_1, t_2, t_3, t_4)^t$ this implies for $i = 1$ that

$$\begin{pmatrix} 0 \\ \frac{1}{2}(d_{22} - 1) \\ 2t_3 + \frac{1}{2}(d_{33} - 1) \\ 2t_4 \end{pmatrix} \in \mathbb{Z}_4$$

and for $i = 2$ that

$$\begin{pmatrix} 0 \\ 2t_2 \\ 2t_3 \\ (d_{44} - 1)\frac{1}{2} \end{pmatrix} \in \mathbb{Z}_4.$$

So d_{22} and d_{44} must be odd integers and since $2t_3$ is an integer the same must hold for d_{33}. To show that d_{11} must be an integer we have to look at $\theta(e_1, I_4)$. Suppose that $\theta(e_1, I_4) = (b, B)$, then we know from the fact that the first entry of the translational parts a_1 and a_2 is zero that the first entry of b must be an integer. Analogously calculations as the one above, imply then that d_{11} also must be an integer.

The 4 other possible linear parts are excluded since we always obtain that some zero entry must also be an odd integer. Let us again demonstrate this, for instance for $\mathfrak{D}_3$. In that case we assume that $\theta(a_2, A_2) = (z_2 + a_2, A_2)$ with $z_2 \in \mathbb{Z}_4$. In the same way as above this implies that

$$\begin{pmatrix} 0 \\ 2t_2 \\ 2t_3 \\ \frac{-1}{2} \end{pmatrix} \in \mathbb{Z}_4.$$

which clearly gives us a contradiction.

Under the given conditions for the possible linear parts one can verify, using for instance the Lemmas 6.8, 6.9 and of course Theorem 2.5 that the Anosov theorem holds for M_1.

Let us now look at the second Bieberbach group E_2 of 22.1.1 by using

the above representation but now we replace a_1 by $\begin{pmatrix} \frac{1}{2} \\ 0 \\ 0 \\ 0 \end{pmatrix}$. Note that

$M_2 = E_2 \backslash \mathbb{R}^4$ is in the same $\mathbb{Z}$-class as M_1. However, in opposite to the results for M_1, we are able to construct a map on M_2 which does not satisfy the Anosov relation. Namely let $f_2 : M_2 \to M_2$ be induced by

$$(\delta, \mathfrak{D}) = (\begin{pmatrix} 0 \\ 0 \\ 0 \\ 0 \end{pmatrix}, \begin{pmatrix} 3\,0\,0\,0 \\ 0\,0\,0\,0 \\ 0\,0\,3\,0 \\ 0\,0\,0\,3 \end{pmatrix}).$$

As before one can verify that this affine endomorphism indeed induces a map on M_2. The explanation for this example is quite simple, because for M_2 we no longer have the condition that d_{22} has to be an odd integer. This comes from the fact that in this case the second entry of a_1 is now also equal to 0. Therefore we can take d_{22} equal to zero and so we obtain that $\det(I_4 - \mathfrak{D}) = -8$ and $\det(I_4 - A_1\mathfrak{D}) = 16$ which by Theorem 2.5 implies that the Anosov relation does not hold for f_2.

A similar example works for the third orientable manifold M_3 obtained by using the third Bieberbach group of 22.1.1. Namely let $f_3 : M_3 \to M_3$ be induced by

$$(\delta, \mathfrak{D}) = (\begin{pmatrix} 0 \\ 0 \\ 0 \\ 0 \end{pmatrix}, \begin{pmatrix} 3\,0\,0\,0 \\ 0\,3\,0\,0 \\ 0\,0\,0\,0 \\ 0\,0\,0\,3 \end{pmatrix}).$$

Another interesting example is the fourth Bieberbach group of 22.1.1.

Replacing in the above representation a_1 by $\begin{pmatrix} \frac{1}{2} \\ \frac{1}{2} \\ \frac{1}{2} \\ 0 \end{pmatrix}$ gives us a Bieber-

bach group E_4 and another manifold $M_4 = E_4 \backslash \mathbb{R}^4$ of the same $\mathbb{Z}$-class. For this manifold we again have that $\mathfrak{D}_7, \mathfrak{D}_8, \mathfrak{D}_{10}, \mathfrak{D}_{12}, \mathfrak{D}_{14}$ and $\mathfrak{D}_{15}$ are suitable linear parts but by the same reasons as for the first manifold we can not use them to construct a counterexample for the Anosov relation.

Now however $\mathfrak{D}_5$ and $\mathfrak{D}_{13}$ are also suitable linear parts under the restrictions that the entries of the first column are odd integers and the entries of the last column are even integers. Let $f_4 : M_4 \to M_4$ be induced by

$$(\delta, \mathfrak{D}) = (\begin{pmatrix} 0 \\ 0 \\ \frac{1}{4} \\ 0 \end{pmatrix}, \begin{pmatrix} 3\,0\,0\,0 \\ 3\,0\,0\,0 \\ 0\,0\,0\,0 \\ 0\,0\,0\,2 \end{pmatrix}),$$

then the Anosov relation does not hold for f_4 since $\det(I_4 - \mathfrak{D}) = 2$ and $\det(I_4 - A_1\mathfrak{D}) = -6$. Note that it is impossible to find a counter example in which the δ-part is equal to zero.

This example shows that to construct counter examples we can not always use affine endomorphisms with a linear part of the form $\mathfrak{D}_7$. To be precise, from Theorem 1.10 we know that for every affine endomorphism $(\delta, \mathfrak{D})$ there exists a unique morphism θ on the universal covering group. In every counterexample up till now, we always used a θ which maps every element of the fundamental group to itself modulo a translation. Amongst other things this implies for every element (a, A) of the fundamental group that $A\mathfrak{D} = \mathfrak{D}A$. Or equivalently that the associated matrices commute and this simplifies a lot the calculations that need to be done. So a possible question could be: if there exist a counterexample, can one always find then a counterexample such that the matrices of the linear parts of the elements of the fundamental group commute with the linear part of the affine endomorphism? M_4 answers this question negatively.

For the other 5 manifolds of this class similar calculations show that the Anosov theorem holds for the 3 manifolds corresponding to the Bieberbach groups of respectively 22.1.4, 22.1.8 and 22.1.12. The counterexamples for the 2 other manifolds are presented below. The Anosov relation does not hold for the map f_5 on the manifold corresponding to the Bieberbach group of 22.1.2 and induced by

$$(\delta, \mathfrak{D}) = (\begin{pmatrix} 0 \\ 0 \\ 0 \\ 0 \end{pmatrix}, \begin{pmatrix} 3\,0\,0\,0 \\ 0\,3\,0\,0 \\ 0\,0\,2\,2 \\ 0\,0\,2\,2 \end{pmatrix}).$$

Similarly for the map f_6 on the manifold corresponding to the Bieberbach group of 22.1.5 and induced by

$$(\delta, \mathfrak{D}) = (\begin{pmatrix} 0 \\ 0 \\ 0 \\ 0 \end{pmatrix}, \begin{pmatrix} 2 & 0 & 0 & 0 \\ 0 & 2 & 2 & -2 \\ 0 & 1 & 1 & -1 \\ 0 & -2 & -2 & 2 \end{pmatrix}).$$

To conclude we may state that this class of manifolds is very interesting and gives us very nice counterexamples which already negatively answers some research questions. Such as for instance: can you always make a counterexample such that the linear part of the homotopy lift commutes with the linear parts of the generators of the fundamental group?

Secondly, it also shows that the translational parts of the elements of the fundamental group can be very crucial for the validity of the Anosov theorem. Except for the results of Chapter 6 this was never the case for the results of the previous chapters.

Finally we proved the following rather surprising result concerning $\mathbb{Z}$-classes of Bieberbach groups (= fundamental groups of flat manifolds).

Proposition 7.2. *There exists flat manifolds M_1 and M_2 such that their fundamental groups belong to the same $\mathbb{Z}$-class and such that the Anosov theorem holds for M_1 but not for M_2.*

7.3 Flat manifolds in dimension 4 with non-abelian holonomy group

While in the previous section the flat manifolds have an abelian holonomy group, this no longer holds for this section. This is interesting since most of our theorems are results concerning manifolds with abelian holonomy group (see chapters 5 and 6). So a next step for future research could be trying to find some extra classes of infra-nilmanifolds with non-abelian holonomy group for which the Anosov theorem holds. In dimension 4 there are 14 of these manifolds, from which there are 10 orientable. The holonomy groups of these 10 manifolds are $S_3, \mathbb{Z}_2 \oplus S_3, D_8$ and A_4.

Moreover, dimension 4 is the lowest dimension in which there exist manifolds having one of the above groups as holonomy group. Now, one could be interested in the fact wether the Anosov theorem holds for infra-nilmanifolds, having a specific holonomy group and appearing in the lowest possible dimension. Or under which extra conditions this could hold. The Klein bottle is a 'counterexample' since dimension 2 is the minimal dimension in which $\mathbb{Z}_2$ occurs as a holonomy group, but nevertheless the Anosov theorem does not hold. On the other hand orientability is for generalized Hantzsche-Wendt manifolds an example of such an extra condition.

For the three flat manifolds with holonomy group S_3 and the one flat manifold with holonomy group $\mathbb{Z}_2 \oplus S_3$ the Anosov theorem does not hold. Since for each of these manifolds we can easily construct a counterexample. For instance let E_1 be the Bieberbach group of 38.1.1 generated by

$$(e_i, I_4), \qquad (a_1, A_1) = (\begin{pmatrix} 0 \\ 0 \\ \frac{1}{2} \\ 0 \end{pmatrix}, \begin{pmatrix} -1 & -1 & 0 & 0 \\ 0 & 1 & 0 & 0 \\ 0 & 0 & 1 & 0 \\ 0 & 0 & 0 & -1 \end{pmatrix})$$

$$\text{and } (a_2, A_2) = (\begin{pmatrix} 0 \\ 0 \\ 0 \\ \frac{2}{3} \end{pmatrix}, \begin{pmatrix} 0 & 1 & 0 & 0 \\ -1 & -1 & 0 & 0 \\ 0 & 0 & 1 & 0 \\ 0 & 0 & 0 & 1 \end{pmatrix})$$

with 1 in the i-th place and 0 everywhere else for e_i; $i = 1, \ldots 4$. Then $M_1 = E_1 \backslash \mathbb{R}^4$ is indeed a flat orientable manifold with S_3 as holonomy group. Let $f_1 : M_1 \to M_1$ be induced by

$$(\delta, \mathfrak{D}) = (\begin{pmatrix} 0 \\ 0 \\ 0 \\ 0 \end{pmatrix}, \begin{pmatrix} 0 & 0 & 0 & 0 \\ 0 & 0 & 0 & 0 \\ 0 & 0 & 3 & 0 \\ 0 & 0 & 0 & 4 \end{pmatrix}).$$

Then $(\delta, \mathfrak{D})$ indeed induces a map on M_1 and $\det(I_4 - \mathfrak{D}) = 6$ and $\det(I_4 - A_2\mathfrak{D}) = -10$. The same example works if we replace E_1 by the Bieberbach group of 38.2.1.

Analogous results holds for the Bieberbach group of 38.2.3 with

$$(\delta, \mathfrak{D}) = (\begin{pmatrix} 0 \\ 0 \\ 0 \\ 0 \end{pmatrix}, \begin{pmatrix} 4 & 0 & 0 & 0 \\ 0 & 1 & 1 & -1 \\ 0 & 1 & 1 & -1 \\ 0 & -1 & -1 & 1 \end{pmatrix})$$

and the Bieberbach group of 81.1.1 with

$$(\delta, \mathfrak{D}) = (\begin{pmatrix} 0 \\ 0 \\ 0 \\ 0 \end{pmatrix}, \begin{pmatrix} 0\,0\,0\,0 \\ 0\,0\,0\,0 \\ 0\,0\,7\,0 \\ 0\,0\,0\,3 \end{pmatrix}).$$

On the other hand for the two flat manifolds with A_4 as holonomy group the Anosov theorem holds. To prove this statement we need some notations. Let E_2 be the Bieberbach group of 44.2.1 generated by

$$(e_i, I_4), \qquad (b_1, B_1) = (\begin{pmatrix} 0 \\ 0 \\ 0 \\ \frac{2}{3} \end{pmatrix}, \begin{pmatrix} 0 & 0 & 1 & 0 \\ -1 & 0 & 0 & 0 \\ 0 & -1 & 0 & 0 \\ 0 & 0 & 0 & 1 \end{pmatrix})$$

$$\text{and } (b_2, B_2) = (\begin{pmatrix} \frac{1}{2} \\ \frac{1}{2} \\ 0 \\ 0 \end{pmatrix}, \begin{pmatrix} -1 & 0 & 0 & 0 \\ 0 & 1 & 0 & 0 \\ 0 & 0 & -1 & 0 \\ 0 & 0 & 0 & 1 \end{pmatrix})$$

and E_3 the Bieberbach group of 44.1.2 generated by

$$(e_i, I_4), \qquad (c_1, C_1) = (\begin{pmatrix} \frac{1}{3} \\ 0 \\ \frac{1}{3} \\ 0 \end{pmatrix}, \begin{pmatrix} 0 & -1 & 1 & 1 \\ -1 & -1 & 1 & 1 \\ 0 & -1 & 1 & 0 \\ 0 & -1 & 0 & 1 \end{pmatrix})$$

$$\text{and } (c_2, C_2) = (\begin{pmatrix} \frac{5}{6} \\ \frac{1}{2} \\ 0 \\ 0 \end{pmatrix}, \begin{pmatrix} -1 & 0 & 0 & 0 \\ 0 & 1 & -1 & -1 \\ 0 & 0 & 0 & -1 \\ 0 & 0 & -1 & 0 \end{pmatrix})$$

with 1 in the i-th place and 0 everywhere else for e_i; $i = 1, \ldots 4$.
Then $M_2 = E_2 \backslash \mathbb{R}^4$ and $M_3 = E_3 \backslash \mathbb{R}^4$ are the manifolds under study. To verify our statement we work very similar to the previous section. Namely we start with calculating all possible linear parts for M_2 but now we have to consider 12 possibilities for as well B_1 as B_2 which result in 144 possible matrices. However 120 of these matrices are of the form

$$\begin{pmatrix} 0\,0\,0\,d_{14} \\ 0\,0\,0\,d_{24} \\ 0\,0\,0\,d_{34} \\ 0\,0\,0\,d_{44} \end{pmatrix}$$

with $d_{ij} \in \mathbb{Z}$. For instance, consider the case where $\mathfrak{D}B_1 = \mathfrak{D}$ and $B_1\mathfrak{D} = \mathfrak{D}B_2$, then

$$\mathfrak{D} = \begin{pmatrix} 0\,0\,0 & d_{14} \\ 0\,0\,0 & -d_{14} \\ 0\,0\,0 & d_{14} \\ 0\,0\,0 & d_{44} \end{pmatrix}$$

Note that this example also shows that there could be some extra restrictions on the d_{j4}. Using Theorem 2.5 we easily obtain that maps induced on M_2 by affine endomorphisms with a linear part of the above form always satisfy the Anosov relation. Note that the same holds for M_3 since we have that $C_1 = PB_1P^{-1}$ and $C_2 = PB_2P^{-1}$ with

$$P = \begin{pmatrix} 1\,0\,1 & 0 \\ 0\,1\,1 & 0 \\ 0\,0\,1 & -1 \\ 0\,0\,1 & 1 \end{pmatrix}.$$

So again we only have to calculate the possible linear parts once. The other 24 matrices are of the form $\begin{pmatrix} D & 0 \\ 0 & d_{44} \end{pmatrix}$ with D a (3×3)-matrix with in each row and each column exactly one non-zero element. For instance, consider the case where $B_1^2 B_2 \mathfrak{D} = \mathfrak{D}B_2$ and $B_2\mathfrak{D} = \mathfrak{D}B_2$, then

$$\mathfrak{D} = \begin{pmatrix} 0 & 0 & d_{13} & 0 \\ 0 & d_{13} & 0 & 0 \\ -d_{13} & 0 & 0 & 0 \\ 0 & 0 & 0 & d_{44} \end{pmatrix}.$$

Moreover, what is demonstrated in this example happens in any case. Namely, suppose that in the first row k appears, then in the other rows only $\pm k$ can appear. For instance D can be

$$\begin{pmatrix} 0\,k & 0 \\ 0\,0 & -k \\ k\,0 & 0 \end{pmatrix} \qquad \text{or} \qquad \begin{pmatrix} 0\,k\,0 \\ k\,0\,0 \\ 0\,0\,k \end{pmatrix}.$$

If we then calculate the determinants of Theorem 2.5 for each of these possible linear parts, we always obtain that the 12 determinants are of the form

$$(1 - d_{44})(1 - k)X \qquad\qquad (7.1)$$

with $X \geq 0$. For instance if we use our example once more we find two possible values for the determinants: $(1 - d_{44})(1 - d_{13})(1 + d_{13}^2)$ and $(1 - d_{44})(1 - d_{13})(1 + d_{13})^2$.

The equation in 7.1 implies that the Anosov relation must hold for the maps on M_2 induced by affine endomorphisms with such a linear part since clearly all these determinants have the same sign.

Again, because of P the same holds for M_3 since, as is demonstrated before, P has 'no influence' on the sign of the determinants in Theorem 2.5.

It is important to note that we do not have to analyze the second type of conditions which come from the translational parts since we already know that the Anosov relation holds. However analyzing these conditions can, as is demonstrated in the previous section, possibly reduce the number of suitable linear parts. We did not verify this since we already had an answer to our question.

Finally for the 4 flat manifolds with D_8 as holonomy group the situation is very similar to the case of the 4-dimensional flat manifolds with $\mathbb{Z}_2 \oplus \mathbb{Z}_2$ as holonomy group. Namely, for the infra-nilmanifolds obtained from the first and the third Bieberbach group of 29.1.1 we can construct a counterexample. These counterexamples are respectively induced by

$$(\delta, \mathfrak{D}) = (\begin{pmatrix} 0 \\ 0 \\ 0 \\ 0 \end{pmatrix}, \begin{pmatrix} 0 & 0 & 0 & 0 \\ 0 & 0 & 0 & 0 \\ 0 & 0 & 5 & 0 \\ 0 & 0 & 0 & 5 \end{pmatrix})$$

and by

$$(\delta, \mathfrak{D}) = (\begin{pmatrix} 0 \\ 0 \\ 0 \\ \frac{1}{8} \end{pmatrix}, \begin{pmatrix} 0 & 0 & 1 & 0 \\ 0 & 0 & 0 & 0 \\ 0 & 0 & 3 & 0 \\ 0 & 0 & 0 & 4 \end{pmatrix}).$$

Note that for the second counterexample, the same holds as in the case of the fourth Bieberbach group of 22.1.1. Namely again, we are not able to use an affine endomorphism for which the linear part commutes with the linear parts of the generators of the fundamental group.

For the manifolds corresponding to the second Bieberbach group of 29.1.1 and the Bieberbach group of 29.1.2 we found by similar calculations as before that the Anosov theorem does hold. So again we found 2 manifolds from the same $\mathbb{Z}$-class for which the Anosov theorem only holds in one case. Therefore the observations of the previous sections are confirmed by these manifolds.

To conclude we may state that these 10 manifolds were very interesting to investigate. They showed for instance that the minimal dimension need not to be restrictive (see S_3), that similar things may happen as in the case of abelian holonomy groups (see D_8) and most interestingly we perhaps found the top of the iceberg of a new class of manifolds for which the Anosov theorem holds (see A_4). Which leads to the following question.

Question 7.3. *Does the Anosov theorem hold for any (flat) manifold with holonomy group A_4?*

Chapter 8

Infra-nilmanifolds

In principle, for infra-nilmanifolds we have to do the same calculations as for flat manifolds. Namely, again by the methods described before, we have to calculate all possible affine endomorphisms and verify Theorem 2.5. However there are two complications. First of all the known descriptions of the almost-Bieberbach groups in low dimensions are not as 'nice' as in the flat case. E.g. from the given group presentation in [12] one can not readily obtain the needed translational and linear parts of a given element. Secondly, the fact that we are now working with non-abelian Lie groups, implies that there are extra restrictions on the linear parts of the suitable affine endomorphisms.

Therefore we start this chapter by formally showing how the calculations can be carried out for 4-dimensional 2-step nilpotent infra-nilmanifolds. The 3-dimensional case is completely similar but less complicated. We also briefly indicate how we deal with the 4-dimensional, 3-step nilpotent case.

In the following 3 sections we then summarize the results and as before we present all the counter examples and give the proofs of our statements.

As in the previous chapter, we need to be able to refer explicitly to specific infra-nilmanifolds. Since an infra-nilmanifold is also completely determined by its fundamental group, we now need a classification of the almost-Bieberbach groups. Such a classification is available in [12], where the almost-Bieberbach groups are divided into numbered families, each still depending on parameters k_i. In dimension 3 there are at most 4 of these parameters, while in dimension 4 there are at most 7. For each family, the possible parameters are specified and a presentation of each almost-Bieberbach group belonging to this family is given. Besides this presentation, also a matrix representation λ of the almost-Bieberbach group is given. For example, for the second family in dimension 3 we find the following entry ([12, p.160]).

$$E = \langle a, b, c, \alpha \mid [b, a] = c^{k_1}, [c, a] = 1, [c, b] = 1,$$
$$\alpha c = c\alpha, \alpha a = a^{-1}\alpha c^{k_2}, \alpha b = b^{-1}\alpha c^{k_3},$$
$$\alpha^2 = c^{k_4}\rangle$$

$$\lambda(\alpha) = \begin{pmatrix} 1 & k_2 & k_3 & \frac{k_4}{2} \\ 0 & -1 & 0 & 0 \\ 0 & 0 & -1 & 0 \\ 0 & 0 & 0 & 1 \end{pmatrix}$$

AB-groups: $k > 0, k \equiv 0 \bmod 2, E = \langle (k, 0, 0, 1) \rangle$

For any possible choice of an integral value of the parameters k_i, the presentation given above, determines an almost-crystallographic group, generated by four elements, a, b, c and α. There is no full information on possible isomorphisms between almost-crystallographic groups determined by different choices of the parameters. However, full information is available for the subset of almost-Bieberbach groups, the torsion free almost-crystallographic groups.

Indeed, the possible almost-Bieberbach groups E are given by the condition on the parameters denoted in the last line. In the case above, this states that $k_1 = k$ must be a positive and even integer, that $k_2 = k_3 = 0$ and that $k_4 = 1$. Note that other choices of parameters can also lead to an almost-Bieberbach group, but then this group will be isomorphic to exactly one with a set of parameters as listed in the last line of the data above.

For any of the almost-crystallographic groups determined by a presentation as above, the a, b, c always generate the Fitting subgroup N (so the lattice in the corresponding Lie group) and the restriction of the matrix representation to this Fitting subgroup can be described in a general form for all the almost-crystallographic groups in dimension 3 at once. Namely, in all cases we have that $[b, a] = c^l, [c, a] = 1, [c, b] = 1$ for some integer l and then the matrix representation λ is given by

$$\lambda(a) = \begin{pmatrix} 1 & 0 & \frac{-l}{2} & 0 \\ 0 & 1 & 0 & 1 \\ 0 & 0 & 1 & 0 \\ 0 & 0 & 0 & 1 \end{pmatrix} \quad \lambda(b) = \begin{pmatrix} 1 & \frac{l}{2} & 0 & 0 \\ 0 & 1 & 0 & 0 \\ 0 & 0 & 1 & 1 \\ 0 & 0 & 0 & 1 \end{pmatrix} \quad \lambda(c) = \begin{pmatrix} 1 & 0 & 0 & 1 \\ 0 & 1 & 0 & 0 \\ 0 & 0 & 1 & 0 \\ 0 & 0 & 0 & 1 \end{pmatrix}$$

(Note that in our example above $l = k_1$). This representation of the Fitting subgroup extends to a representation of the full almost-crystallographic group by defining the images of generators as indicated in the entry above. We will explain in the next section, how this affine representation really contains all the information we need.

In the 4-dimensional case analogous entries can be found, but they are slightly more complicated since for instance now four generators of the Fitting subgroup are given and there are more relations between them. Moreover, we also have to consider 3-step nilpotent Lie groups.

Finally, note that during this chapter, G always denotes the Lie group corresponding to the almost-Bieberbach group E under study (or the universal covering group of the infra-nilmanifold) and N refers to the Fitting subgroup in E (which is also the corresponding lattice in G).

8.1 Calculations on 4 dimensional infra-nilmanifolds

In this section we shortly explain how we are able to make the necessary calculations with the almost Bieberbach groups we need to investigate. We first present a construction which can be used in general for the 2-step nilpotent case, but not for higher nilpotency classes. Afterwards we give, in a second section, a specific method, which we can use in the 4-dimensional 3-step nilpotent case.

8.1.1 2-step nilpotent infra-nilmanifolds

In this section we demonstrate how we can use the affine representation of the fundamental group to obtain the information that we need. We do this in the (more general) case where the universal covering group is a 4-dimensional, 2-step nilpotent Lie group. Therefore, let $\mathfrak{g}$ be the 4-dimensional, 2-step nilpotent Lie algebra determined by

$$\mathfrak{g} = \langle A, B, C, D \mid [B, A] = l_1 D, [C, A] = l_2 D, [C, B] = l_3 D,$$
$$[D, A] = 0, [D, B] = 0, [D, C] = 0 \rangle$$

Here, $l_1, l_2, l_3 \in \mathbb{Z}$. (In fact, any non-zero choice of parameters actually determines, up to some isomorphism, the same Lie algebra, but this is not important to what follows).

Note that in this chapter we use the notations of [12] and we refer to this book for more details. This algebra has a faithful matrix representation:

$$\varphi : \mathfrak{g} \to M_5(\mathbb{R}) : xA+yB+zC+tD \mapsto \begin{pmatrix} 0 & \frac{l_1y+l_2z}{2} & \frac{-l_1x+l_3z}{2} & \frac{-l_1x-l_2y}{2} & t \\ 0 & 0 & 0 & 0 & x \\ 0 & 0 & 0 & 0 & y \\ 0 & 0 & 0 & 0 & z \\ 0 & 0 & 0 & 0 & 0 \end{pmatrix}.$$

Then $G = \exp(\mathfrak{g})$ is a connected, simply connected, 2-step nilpotent Lie group with Lie algebra $\mathfrak{g}$ and a faithful matrix representation is given by

$$\psi : G \to \mathrm{Gl}(5,\mathbb{R}) : \exp(xA+yB+zC+tD) \mapsto \exp(\varphi(xA+yB+zC+tD))$$

Let $a = \exp(A), b = \exp(B), c = \exp(C), d = \exp(D)$, and let N be the subgroup of G, generated by a, b, c and d. One can check that N has a presentation:

$$N = \langle a, b, c, d \mid [b, a] = d^{l_1}, [c, a] = d^{l_2}, [c, b] = d^{l_3},$$
$$[d, a] = 1, [d, b] = 1, [d, c] = 1\rangle.$$

Later on, we need to able to work with z^r for any $z \in N, r \in \mathbb{R}$, which is obtained as follows $z^r = \exp(r \log z)$.

In [12, p 168], we find a matrix representation $\lambda : N \to \mathrm{Gl}(5,\mathbb{R})$ for N, given by

$$\lambda(a) = \begin{pmatrix} 1 & 0 & \frac{-l_1}{2} & \frac{-l_2}{2} & 0 \\ 0 & 1 & 0 & 0 & 1 \\ 0 & 0 & 1 & 0 & 0 \\ 0 & 0 & 0 & 1 & 0 \\ 0 & 0 & 0 & 0 & 1 \end{pmatrix}, \lambda(b) = \begin{pmatrix} 1 & \frac{l_1}{2} & 0 & \frac{-l_3}{2} & 0 \\ 0 & 1 & 0 & 0 & 0 \\ 0 & 0 & 1 & 0 & 1 \\ 0 & 0 & 0 & 1 & 0 \\ 0 & 0 & 0 & 0 & 1 \end{pmatrix},$$

$$\lambda(c) = \begin{pmatrix} 1 & \frac{l_2}{2} & \frac{l_3}{2} & 0 & 0 \\ 0 & 1 & 0 & 0 & 0 \\ 0 & 0 & 1 & 0 & 0 \\ 0 & 0 & 0 & 1 & 1 \\ 0 & 0 & 0 & 0 & 1 \end{pmatrix} \text{ and } \lambda(d) = \begin{pmatrix} 1 & 0 & 0 & 0 & 1 \\ 0 & 1 & 0 & 0 & 0 \\ 0 & 0 & 1 & 0 & 0 \\ 0 & 0 & 0 & 1 & 0 \\ 0 & 0 & 0 & 0 & 1 \end{pmatrix}.$$

It is obvious that λ is just the restriction of ψ to N. Moreover, N is a lattice of G (or $\lambda(N)$ is a lattice of $\psi(G)$) and thus G is the Mal'cev completion of N.

Now we want to extend the above matrix representation ψ, to obtain a matrix representation of $G \rtimes \mathrm{Endo}(G)$. Therefore take $\mathfrak{D} \in \mathrm{Endo}(G)$, then $\mathfrak{D}_*$ is also an element of $\mathrm{Endo}(\mathfrak{g})$ and $\mathfrak{D}_*$ is described by a matrix of the form

$$
T_{\mathfrak{D}_*} = \begin{pmatrix} \epsilon & p_1 & q_1 & r_1 \\ 0 & p_2 & q_2 & r_2 \\ 0 & p_3 & q_3 & r_3 \\ 0 & p_4 & q_4 & r_4 \end{pmatrix}
\tag{8.1}
$$

with respect to the basis D, A, B, C. Moreover ϵ must satisfy certain, depending on the values of the parameters l_i, of the following relations:

$$\epsilon l_1 = l_1(p_2 q_3 - q_2 p_3) + l_2(p_2 q_4 - q_2 p_4) + l_3(p_3 q_4 - q_3 p_4)$$
$$\epsilon l_2 = l_1(p_2 r_3 - r_2 p_3) + l_2(p_2 r_4 - r_2 p_4) + l_3(p_3 r_4 - r_3 p_4)$$
$$\epsilon l_3 = l_1(q_2 r_3 - r_2 q_3) + l_2(q_2 r_4 - r_2 q_4) + l_3(q_3 r_4 - r_3 q_4)$$

The ith equation has to be satisfied if and only if $l_i \neq 0$.

Note that also the converse holds: any matrix of the above form is the matrix of the differential of an element of $\mathrm{Endo}(G)$. An easy calculation, using (8.1), then shows that

$$
\chi : G \rtimes \mathrm{Endo}(G) \to M_5(\mathbb{R}) : (\delta, \mathfrak{D}) \mapsto \psi(\delta) \begin{pmatrix} T_{\mathfrak{D}_*} & 0 \\ 0 & 1 \end{pmatrix}
$$

is a faithful representation of the semi-group $G \rtimes \mathrm{Endo}(G)$ and note that $\chi(G \rtimes \mathrm{Aut}(G)) = \chi(G \rtimes \mathrm{Endo}(G)) \cap \mathrm{Gl}(5, \mathbb{R})$. Obviously χ extends ψ and λ.

If we now consider any 4-dimensional, 2-step nilpotent Bieberbach group E as listed in [12], then it is clear that $\lambda(E) \subseteq \chi(G \rtimes \mathrm{Aut}(G))$. This implies that, using the faithful representations λ and χ, we can view E as a discrete subgroup of $G \rtimes \mathrm{Aut}(G)$ such that $G \cap E = N$ and E/N is finite. Using this construction and the matrices of [12], we can now easily make calculations in $G \rtimes \mathrm{Endo}(G)$. To do that, one has to decompose the matrices of the generators into a product of the matrix associated to the translational part and the matrix associated to the linear part. Once we have these matrices, we can proceed analogously to the calculations for the flat manifolds. Let us illustrate this by means of an example using the third family in dimension 4 ([12, p 170].

Example 8.1. *In [12] we find the following entry.*

$$E = \langle a, b, c, d\alpha \mid [b,a] = 1, [c,a] = d^{k_1}, [c,b] = 1,$$
$$\alpha c = c^{-1}\alpha d^{k_3}, \alpha a = a^{-1}\alpha d^{k_2}, \alpha b = b\alpha,$$
$$\alpha d = d\alpha, \alpha^2 = d^{k_4}\rangle$$

$$\lambda(\alpha) = \begin{pmatrix} 1 & k_2 & 0 & k_3 & \frac{1}{2} \\ 0 & -1 & 0 & 0 & 0 \\ 0 & 0 & 1 & 0 & 0 \\ 0 & 0 & 0 & -1 & 0 \\ 0 & 0 & 0 & 0 & 1 \end{pmatrix}$$

AB-groups: $k > 0, k \equiv 0 \bmod 2, E = \langle (k, 0, 0, 1)\rangle$

This implies that for the almost-Bieberbach groups E of this family that $l_1 = l_3 = k_2 = k_3 = 0$, $k_4 = 1$, l_2 is an even integer and

$$\lambda(\alpha) = \begin{pmatrix} 1 & 0 & 0 & 0 & \frac{1}{2} \\ 0 & 1 & 0 & 0 & 0 \\ 0 & 0 & 1 & 0 & 0 \\ 0 & 0 & 0 & 1 & 0 \\ 0 & 0 & 0 & 0 & 1 \end{pmatrix} \begin{pmatrix} 1 & 0 & 0 & 0 & 0 \\ 0 & -1 & 0 & 0 & 0 \\ 0 & 0 & 1 & 0 & 0 \\ 0 & 0 & 0 & -1 & 0 \\ 0 & 0 & 0 & 0 & 1 \end{pmatrix} = \lambda(d^{\frac{1}{2}}) \begin{pmatrix} 1 & 0 & 0 & 0 & 0 \\ 0 & -1 & 0 & 0 & 0 \\ 0 & 0 & 1 & 0 & 0 \\ 0 & 0 & 0 & -1 & 0 \\ 0 & 0 & 0 & 0 & 1 \end{pmatrix}.$$

By the method described above, this implies that the translational part of α is equal to $d^{\frac{1}{2}}$ and that the differential of the linear part is described by the matrix

$$\begin{pmatrix} 1 & 0 & 0 & 0 \\ 0 & -1 & 0 & 0 \\ 0 & 0 & 1 & 0 \\ 0 & 0 & 0 & -1 \end{pmatrix}.$$

From this we may conclude that the infra-nilmanifold which has E as its fundamental group is orientable and has $\mathbb{Z}_2$ as holonomy group. Let $(\delta, \mathfrak{D})$ be an affine endomorphism of G, then $\mathfrak{D}_*$ is represented by a matrix of the form

$$\begin{pmatrix} \epsilon & p_1 & q_1 & r_1 \\ 0 & p_2 & q_2 & r_2 \\ 0 & p_3 & q_3 & r_3 \\ 0 & p_4 & q_4 & r_4 \end{pmatrix}$$

satisfying $\epsilon l_2 = l_2(p_2 r_4 - r_2 p4) \Leftrightarrow \epsilon = p_2 r_4 - r_2 p_4$.

An analogous, but easier construction, can be carried out in the three dimensional case.

8.1.2 3-step nilpotent infra-nilmanifolds

In this section we briefly describe the method we used for the 3-step nilpotent case. Since we restrict ourselves to dimension 4, we find in [12] that we only have to deal with infra-nilmanifolds with holonomy group $\mathbb{Z}_2$ or $\mathbb{Z}_2 \oplus \mathbb{Z}_2$. Moreover in the latter case the manifolds are always non-orientable and therefore we only focus on the infra-nilmanifolds with $\mathbb{Z}_2$ as holonomy group.

Assume that E is a fundamental group of such an infra-nilmanifold and is generated by a, b, c, d, α. Again we want to replace α by an affine map of the form $(a^{t_1} b^{t_2} c^{t_3} d^{t_4}, \mathfrak{A})$. (Note that we are still assuming that G is the Mal'cev completion of the lattice N generated by a, b, c, d and that $\mathfrak{g}$ is the corresponding Lie algebra generated by D, C, A, B.) In the previous section we could avoid the actual construction of $\mathfrak{A}$ by using only $\mathfrak{A}_*$. Now, we will really need $\mathfrak{A}$ in order to construct $\mathfrak{A}_*$. However, the first thing we will do is to determine the translational part of the generator α with a non-trivial linear part. For this, we explicitly use that the order of α is 2. This implies that we can find $t_1, \ldots, t_4$ by using the respective matrix representations of a, b, c, d, α and the fact that

$$\lambda(((a^{t_1} b^{t_2} c^{t_3} d^{t_4})^{-1} \alpha)^2) = I_5$$

Note that in the 3-step nilpotent case the matrix representation of the lattice is not the same for all the almost-Bieberbach groups. In the following we will denote $(a^{t_1} b^{t_2} c^{t_3} d^{t_4})^{-1} \alpha$, with the t_i as found above, by β.

We can now determine $\mathfrak{A}_*$ by explicitly computing the images of D, C, A, B. The image of for instance A is given by

$$\mathfrak{A}_*(A) = \log(\mathfrak{A}(\exp(a)))$$
$$= q_1 D + q_2 C q_3 A +_4 B$$

with $q_i \in \mathbb{R}$ and analogously for the other elements. Now to calculate this we can use the matrix representations, since we have that $\mathfrak{A}(g) = \lambda(\beta g \beta^{-1}) = \lambda(\beta g \beta)$. This allows us to explicitly determine $\mathfrak{A}_*$. Note that because of the structure of G, we have that $\mathfrak{A}_*(D)$ only depends on D and $\mathfrak{A}_*(C)$ only on D, C. To calculate the respective coefficients we need the relations between D, C, A, B and the images of A and B.

Of course the same can be done to calculate the δ and $\mathfrak{D}_*$ of the suitable affine endomorphisms $(\delta, \mathfrak{D})$. Note that since we may assume that $\mathbb{Z}_2$ is the holonomy group, we may because of Proposition 2.6 assume that $\mathfrak{A}\mathfrak{D} = \mathfrak{A}\mathfrak{D}$ and also that

$$(\delta, \mathfrak{D})(a^{t_1}b^{t_2}c^{t_3}d^{t_4}, \mathfrak{A}) = (na^{t_1}b^{t_2}c^{t_3}d^{t_4}, \mathfrak{A})(\delta, \mathfrak{D})$$

with $n \in N$.

8.2 The 3-dimensional, 2-step infra-nilmanifolds

As in the previous chapter we summarize our results in tables, but now the situation is much more delicate because of the presence of the parameters k_1, k_2, k_3, k_4. For a given family the k_2, k_3, k_4 are always fixed integers, but k_1 is a strict positive integer $(k_1 > 0)$ and in most occasions it has to satisfy some extra conditions. In [12] k_1 is replaced by k and we use the same convention. It is clear that for each value of $k_1 > 0$ we have another almost-Bieberbach group and so also another infra-nilmanifold.

Therefore in each dimension, we present for each possible holonomy group F a table consisting of all families of almost-Bieberbach groups which have F as holonomy group. In the first two columns, we present the families and summarize the conditions on k (if there are any). In the following columns we summarize as before all possibilities concerning the Anosov relation for the corresponding manifolds. For the enumeration of the almost-Bieberbach groups and more details we refer to [12].

For the nilmanifolds in dimension 3 we have the following table.

1				
Family	k	Orientable		Non-orientable
		An.th. holds	An.th. holds not	An.th. holds not
1	/	1	0	0

This means that for each $k > 0$ we have a nilmanifold for which of course the Anosov theorem holds.

For the infra-nilmanifolds with $\mathbb{Z}_2$ as holonomy group we already have an interesting result.

Family	k	$\mathbb{Z}_2$		
		Orientable		Non-orientable
		An.th. holds	An.th. holds not	An.th. holds not
2	$k \equiv 0 \bmod 2$	0	1	0
4	$k \equiv 0 \bmod 2$	1	0	0

Indeed, this table indicates that for each $k > 0$ such that $k \equiv 0 \bmod 2$ we have one infra-nilmanifold for which the Anosov theorem holds (Family 4) and one infra-nilmanifold for which it does not hold (Family 2). Family 2 is another, much more simple, example that in general Proposition 5.12 does not hold for infra-nilmanifolds. To prove this, we use the techniques described in the previous section to find that α can be viewed as $(c^{\frac{1}{2}}, \mathfrak{A})$ such that $\mathfrak{A}_*$ is determined by the matrix

$$T_{\mathfrak{A}_*} = \begin{pmatrix} 1 & 0 & 0 \\ 0 & -1 & 0 \\ 0 & 0 & -1 \end{pmatrix}.$$

Now let f be the map on this infra-nilmanifold M induced by $(1, \mathfrak{D})$, where $\mathfrak{D}_*$ is given by

$$T_{\mathfrak{D}_*} = \begin{pmatrix} -1 & 0 & 0 \\ 0 & 1 & 1 \\ 0 & 2 & 1 \end{pmatrix}.$$

It is easy to verify that $(1, \mathfrak{D})$ really induces a map on M, since, by using the matrices above, one computes that $\mathfrak{D}\mathfrak{A} = \mathfrak{A}\mathfrak{D}$ and hence

$$\begin{aligned} (1, \mathfrak{D})(c^{\frac{1}{2}}, \mathfrak{A}) &= (\mathfrak{D}(c^{\frac{1}{2}}), \mathfrak{D}\mathfrak{A}) \\ &= (c^{\frac{-1}{2}}, \mathfrak{D}\mathfrak{A}) \\ &= (c^{-1}, \mathcal{E})(c^{\frac{1}{2}}, \mathfrak{A}\mathfrak{D}) \\ &= (c^{-1}, \mathcal{E})(c^{\frac{1}{2}}, \mathfrak{A})(1, \mathfrak{D}). \end{aligned}$$

(Here $\mathcal{E}$ denotes the identity endomorphism of G). For this map the Anosov relation does not hold since $\det(I_3 - T_{\mathfrak{D}_*}) = -4$, while $(\det(I_3 - T_{\mathfrak{A}_*}T_{\mathfrak{D}_*}) = 4$.

Note that $\mathfrak{A}$ generates the holonomy group of M and therefore $T_{\mathfrak{A}_*}$ may be seen as the holonomy representation of $\mathfrak{A}$. In the sequel, we will slightly abuse notation by using the familiar $T_*(\mathfrak{A})$ instead of $T_{\mathfrak{A}_*}$ (without formally defining the holonomy representation) and $\mathfrak{D}_*$ instead of $T_{\mathfrak{D}_*}$.

To verify the statement for Family 4, we need a description of all suitable affine endomorphisms $(\delta, \mathfrak{D}) \in G \rtimes \mathrm{Endo}(G)$. For such a suitable affine endomorphism, we have for any $(t, \mathfrak{T})$ that

$$\theta(t, \mathfrak{T})(\delta, \mathfrak{D}) = (\delta, \mathfrak{D})(t, \mathfrak{T}).$$

Supposing that $\theta(t, \mathfrak{T}) = (t_1, \mathfrak{T}_1)$, this implies that

$$(t_1 \mathfrak{T}_1(\delta), \mathfrak{T}_1 \mathfrak{D}) = (\delta \mathfrak{D}(t), \mathfrak{D} \mathfrak{T})$$

and as in the flat case, we obtain the two types of restrictions: one on the linear parts and one on the translational parts. For family 4 there is only one generator with non-trivial linear part, namely $(a^{\frac{1}{2}}, \mathfrak{A})$ with $T_*(\mathfrak{A}) = \begin{pmatrix} -1 & 0 & 0 \\ 0 & 1 & 0 \\ 0 & 0 & -1 \end{pmatrix}$. Because of Proposition 2.6, we only have to consider the $(\delta, \mathfrak{D})$ for which $\mathfrak{D}_*$ commutes with $T_*(\mathfrak{A})$. This implies that

$$\mathfrak{D}_* = \begin{pmatrix} d_{22} d_{33} & 0 & d_{13} \\ 0 & d_{22} & 0 \\ 0 & 0 & d_{33} \end{pmatrix}$$

where all the entries of $\mathfrak{D}_*$ have to be integers. Actually this last claim, follows from the fact that the corresponding homomorphism has to map the lattice N into itself. If this would not be the case, then again because of Proposition 2.6, the Anosov theorem holds and we can forget about this case.

The restriction on the translational part for this case becomes

$$na^{\frac{1}{2}} \mathfrak{A}(\delta) = \delta \mathfrak{D}(a^{\frac{1}{2}})$$

with $n \in N$ (and so $t_1 = na^{\frac{1}{2}}$). Equivalently, we have that

$$\delta \mathfrak{D}(a^{\frac{1}{2}})(a^{\frac{1}{2}} \mathfrak{A}(\delta))^{-1} = \delta \mathfrak{D}(a^{\frac{1}{2}}) \mathfrak{A}(\delta^{-1}) a^{\frac{-1}{2}} \in N.$$

The above statement can be verified by using the faithful matrix representation λ to rewrite the element. Indeed, the statement holds if and only if this matrix representation can be written as $\lambda(a^{z_1} b^{z_2} c^{z_3})$ with $z_1, z_2, z_3 \in \mathbb{Z}$. Taking $\delta = a^{d_1} b^{d_2} c^{d_3}$, we obtain

$$\lambda(\delta \mathfrak{D}(a^{\frac{1}{2}}) \mathfrak{A}(\delta^{-1}) a^{\frac{-1}{2}}) = \begin{pmatrix} 1 & ld_2 & \frac{l}{4}(1 - d_{22}) & -\frac{ld_2}{2}(1 + 4d_1) + 2d_3 \\ 0 & 1 & 0 & \frac{1}{2}(-1 + d_{22}) \\ 0 & 0 & 1 & 2d_2 \\ 0 & 0 & 0 & 1 \end{pmatrix}$$

$$= \lambda(a^{\frac{-1+d_{22}}{2}}) \lambda(b^{2d_2}) \lambda(c^{\frac{ld_2}{2}(-2+d_{22}-4d_1)+2d_3}).$$

So to satisfy the restrictions on the translational parts, one of the implications is that $d_{22} \in 1 + 2\mathbb{Z}$. There are also restrictions on δ, but these will have no influence on the validity of the Anosov theorem and so we disregard them.

If we now calculate the determinants occurring in Theorem 2.5, we obtain that $\det(I_3 - \mathfrak{D}_*) = (1 - d_{22}d_{33})(1 - d_{22})(1 - d_{33})$ and $\det(I_3 - T_*(\mathfrak{A})\mathfrak{D}_*) = (1 + d_{22}d_{33})(1 - d_{22})(1 + d_{33})$. To possibly obtain a counterexample, these determinants should have different signs, so we only have to focus on the first and the last factor. Since d_{22} can not be zero, we can easily verify that $(1 - d_{22}d_{33})(1 - d_{33})$ and $(1 + d_{22}d_{33})(1 + d_{33})$ have the same sign. Indeed, this certainly holds if $d_{33} = 0$ or ± 1 and can be checked if $|d_{33}| > 1$ since $d_{22} \neq 0$. So we may conclude that the Anosov theorem holds for all the manifolds belonging to Family 4.

In the following 3 tables we present the results for the other infra-nilmanifolds with cyclic holonomy group in dimension 3. All these results follow directly from Theorem 5.4.

$\mathbb{Z}_3$				
Family	k	Orientable		Non-orientable
		An.th. holds	An.th. holds not	An.th. holds not
13	$k \equiv 0 \bmod 3$	2	0	0
	$k \not\equiv 0 \bmod 3$	1	0	0

$\mathbb{Z}_4$				
Family	k	Orientable		Non-orientable
		An.th. holds	An.th. holds not	An.th. holds not
10	$k \equiv 0 \bmod 2$	1	0	0
	$k \equiv 0 \bmod 4$	1	0	0

$\mathbb{Z}_6$				
Family	k	Orientable		Non-orientable
		An.th. holds	An.th. holds not	An.th. holds not
16	$k \equiv 0 \bmod 6$	2	0	0
	$k \equiv 2 \bmod 6$	1	0	0
	$k \equiv 4 \bmod 6$	1	0	0

The last family we need to examine, are infra-nilmanifolds with $\mathbb{Z}_2 \oplus \mathbb{Z}_2$ as holonomy group. These manifolds are not covered by Theorem 6.10, which is only dealing with flat manifolds. Intriguingly we also obtain that the Anosov theorem holds.

$\mathbb{Z}_2 \oplus \mathbb{Z}_2$			
Family	k	Orientable		Non-orientable
		An.th. holds	An.th. holds not	An.th. holds not
8	$k \equiv 0 \bmod 4$	1	0	0

To prove this, we first have to describe the generators with non-trivial part of the almost-Bieberbach group of Family 8 as affine maps. These are $(c^{\frac{1}{2}}, \mathfrak{A}_1)$ and $(a^{\frac{1}{2}} b^{\frac{1}{2}} c^{\frac{k}{8}}, \mathfrak{A}_2)$ with

$$T_*(\mathfrak{A}_1) = \begin{pmatrix} 1 & \frac{k}{2} & \frac{-k}{2} \\ 0 & -1 & 0 \\ 0 & 0 & -1 \end{pmatrix} \quad \text{and} \quad T_*(\mathfrak{A}_2) = \begin{pmatrix} -1 & \frac{-k}{2} & 0 \\ 0 & 1 & 0 \\ 0 & 0 & -1 \end{pmatrix}.$$

As argued in the previous chapter, to determine all suitable affine endomorphisms $(\delta, \mathfrak{D})$, we again have to consider 16 possibilities for $\mathfrak{D}$ or equivalently for $\mathfrak{D}_*$. These possibilities for $\mathfrak{D}_*$ are given in the table below.

$\mathfrak{D}_1 = \mathfrak{D}_1 \mathfrak{A}_1$ $\mathfrak{D}_1 = \mathfrak{D}_1 \mathfrak{A}_2$ $(\mathfrak{D}_1)_* = \begin{pmatrix} 0 & 0 & 0 \\ 0 & 0 & 0 \\ 0 & 0 & 0 \end{pmatrix}$	$\mathfrak{D}_2 = \mathfrak{D}_2 \mathfrak{A}_1$ $\mathfrak{A}_1 \mathfrak{D}_2 = \mathfrak{D}_2 \mathfrak{A}_2$ $(\mathfrak{D}_2)_* = \begin{pmatrix} 0 & 0 & 0 \\ 0 & 0 & 0 \\ 0 & 0 & 0 \end{pmatrix}$
$\mathfrak{D}_3 = \mathfrak{D}_3 \mathfrak{A}_1$ $\mathfrak{A}_2 \mathfrak{D}_3 = \mathfrak{D}_3 \mathfrak{A}_2$ $(\mathfrak{D}_3)_* = \begin{pmatrix} 0 & 0 & 0 \\ 0 & 0 & 0 \\ 0 & 0 & 0 \end{pmatrix}$	$\mathfrak{D}_4 = \mathfrak{D}_4 \mathfrak{A}_1$ $\mathfrak{A}_1 \mathfrak{A}_2 \mathfrak{D}_4 = \mathfrak{D}_4 \mathfrak{A}_2$ $(\mathfrak{D}_4)_* = \begin{pmatrix} 0 & 0 & 0 \\ 0 & 0 & 0 \\ 0 & 0 & 0 \end{pmatrix}$
$\mathfrak{A}_1 \mathfrak{D}_5 = \mathfrak{D}_5 \mathfrak{A}_1$ $\mathfrak{D}_5 = \mathfrak{D}_5 \mathfrak{A}_2$ $(\mathfrak{D}_5)_* = \begin{pmatrix} 0 & \frac{k}{4}(d_{32} - d_{22}) & 0 \\ 0 & d_{22} & 0 \\ 0 & d_{32} & 0 \end{pmatrix}$	$\mathfrak{A}_1 \mathfrak{D}_6 = \mathfrak{D}_6 \mathfrak{A}_1$ $\mathfrak{A}_1 \mathfrak{D}_6 = \mathfrak{D}_6 \mathfrak{A}_2$ $(\mathfrak{D}_6)_* = \begin{pmatrix} 0 & 0 & \frac{k}{4}(d_{33} - d_{23}) \\ 0 & 0 & d_{23} \\ 0 & 0 & d_{33} \end{pmatrix}$

$$\mathfrak{A}_1\mathfrak{D}_7 = \mathfrak{D}_7\mathfrak{A}_1$$
$$\mathfrak{A}_2\mathfrak{D}_7 = \mathfrak{D}_7\mathfrak{A}_2$$

$$(\mathfrak{D}_7)_* =$$
$$\begin{pmatrix} d_{22}d_{33} & \frac{k}{4}d_{22}(d_{33}-1) & \frac{k}{4}d_{33}(d_{22}-1) \\ 0 & d_{22} & 0 \\ 0 & 0 & d_{33} \end{pmatrix}$$

$$\mathfrak{A}_1\mathfrak{D}_8 = \mathfrak{D}_8\mathfrak{A}_1$$
$$\mathfrak{A}_1\mathfrak{A}_2\mathfrak{D}_8 = \mathfrak{D}_8\mathfrak{A}_2$$

$$(\mathfrak{D}_8)_* =$$
$$\begin{pmatrix} -d_{23}d_{32} & \frac{k}{4}d_{32}(1-d_{23}) & \frac{k}{4}d_{23}(d_{32}-1) \\ 0 & 0 & d_{23} \\ 0 & d_{32} & 0 \end{pmatrix}$$

$\mathfrak{A}_2\mathfrak{D}_9 = \mathfrak{D}_9\mathfrak{A}_1$ $\mathfrak{D}_9 = \mathfrak{D}_9\mathfrak{A}_2$ $$(\mathfrak{D}_9)_* = \begin{pmatrix} 0 & d_{12} & 0 \\ 0 & 0 & 0 \\ 0 & d_{32} & 0 \end{pmatrix}$$	$\mathfrak{A}_2\mathfrak{D}_{10} = \mathfrak{D}_{10}\mathfrak{A}_1$ $\mathfrak{A}_1\mathfrak{D}_{10} = \mathfrak{D}_{10}\mathfrak{A}_2$ $$(\mathfrak{D}_{10})_* = \begin{pmatrix} 0 & d_{12} & \frac{k}{4}d_{33} \\ 0 & 0 & 0 \\ 0 & 0 & d_{33} \end{pmatrix}$$
$\mathfrak{A}_2\mathfrak{D}_{11} = \mathfrak{D}_{11}\mathfrak{A}_1$ $\mathfrak{A}_2\mathfrak{D}_{11} = \mathfrak{D}_{11}\mathfrak{A}_2$ $$(\mathfrak{D}_{11})_* = \begin{pmatrix} 0 & 0 & d_{13} \\ 0 & 0 & 0 \\ 0 & 0 & d_{33} \end{pmatrix}$$	$\mathfrak{A}_2\mathfrak{D}_{12} = \mathfrak{D}_{12}\mathfrak{A}_1$ $\mathfrak{A}_1\mathfrak{A}_2\mathfrak{D}_{12} = \mathfrak{D}_{12}\mathfrak{A}_2$ $$(\mathfrak{D}_{12})_* = \begin{pmatrix} 0 & \frac{k}{4}d_{32} & d_{13} \\ 0 & 0 & d_{23} \\ 0 & d_{32} & 0 \end{pmatrix}$$
$\mathfrak{A}_1\mathfrak{A}_2\mathfrak{D}_{13} = \mathfrak{D}_{13}\mathfrak{A}_1$ $\mathfrak{D}_{13} = \mathfrak{D}_{13}\mathfrak{A}_2$ $$(\mathfrak{D}_{13})_* = \begin{pmatrix} 0 & d_{12} & 0 \\ 0 & d_{22} & 0 \\ 0 & 0 & 0 \end{pmatrix}$$	$\mathfrak{A}_1\mathfrak{A}_2\mathfrak{D}_{14} = \mathfrak{D}_{14}\mathfrak{A}_1$ $\mathfrak{A}_1\mathfrak{D}_{14} = \mathfrak{D}_{14}\mathfrak{A}_2$ $$(\mathfrak{D}_{14})_* = \begin{pmatrix} 0 & d_{12} & \frac{-k}{4}d_{23} \\ 0 & 0 & d_{23} \\ 0 & 0 & 0 \end{pmatrix}$$

$\mathfrak{A}_1\mathfrak{A}_2\mathfrak{D}_{15} = \mathfrak{D}_{15}\mathfrak{A}_1$ $\mathfrak{A}_2\mathfrak{D}_{15} = \mathfrak{D}_{15}\mathfrak{A}_2$ $(\mathfrak{D}_{15})_* = \begin{pmatrix} 0 & 0 & d_{13} \\ 0 & d_{22} & 0 \\ 0 & 0 & 0 \end{pmatrix}$	$\mathfrak{A}_1\mathfrak{A}_2\mathfrak{D}_{16} = \mathfrak{D}_{16}\mathfrak{A}_1$ $\mathfrak{A}_1\mathfrak{A}_2\mathfrak{D}_{16} = \mathfrak{D}_{16}\mathfrak{A}_2$ $(\mathfrak{D}_{16})_* = \begin{pmatrix} 0 & 0 & d_{13} \\ 0 & 0 & d_{23} \\ 0 & 0 & 0 \end{pmatrix}$

Before we analyze the restrictions on the translational parts, we can already drop the cases $\mathfrak{D}_1, \mathfrak{D}_2, \mathfrak{D}_3, \mathfrak{D}_4, \mathfrak{D}_9, \mathfrak{D}_{14}, \mathfrak{D}_{16}$ since for these cases the determinants of Theorem 2.5 are all equal to 1. Secondly, because of Proposition 2.6 we know that the Anosov relation holds for the case of $\mathfrak{D}_{12}$. This reduces our investigation to 8 possibilities. Thirdly, in the case $\mathfrak{D}_8$, the determinants of Theorem 2.5 are all equal to $(1 - d_{23}d_{32})^2 + (1 + d_{23}d_{32})^2$. So again the Anosov theorem holds since all these determinants are strictly positive. So, only 7 cases remain.

From these 7 cases we can eliminate 6 more since they are not compatible with the respective restrictions on the translational parts. For example for $\mathfrak{D}_9$, we have that $\delta\mathfrak{D}(c^{\frac{1}{2}})\mathfrak{A}_2(\delta^{-1})(a^{\frac{1}{2}}b^{\frac{1}{2}}c^{\frac{k}{8}})^{-1}$ is not an element of N since we can rewrite this element as

$$a^{\frac{-1}{2}}b^{\frac{-1}{2}+2d_2}c^{\frac{k}{8}(1+4d_1-16d_1d_2-8d_2+16d_3)}.$$

Note that again we have taken $\delta = (d_1, d_2, d_3)^t$. Similar arguments hold for $\mathfrak{D}_5, \mathfrak{D}_6, \mathfrak{D}_{10}, \mathfrak{D}_{11}, \mathfrak{D}_{13}$ and $\mathfrak{D}_{15}$.

So to check the validity of the Anosov theorem, we only have to focus on the case $\mathfrak{D}_7$. From analyzing the translational restrictions in the way explained before, we obtain that d_{22} and d_{33} must be odd integers. This is very useful since the 4 determinants we have to consider are

$$\det(I_3 - (\mathfrak{D}_7)_*) = (1 - d_{22}d_{33})(1 - d_{22})(1 - d_{33}),$$
$$\det(I_3 - T_*(\mathfrak{A}_1)(\mathfrak{D}_7)_*) = (1 - d_{22}d_{33})(-1 - d_{22})(-1 - d_{33}),$$
$$\det(I_3 - T_*(\mathfrak{A}_2)(\mathfrak{D}_7)_*) = (-1 - d_{22}d_{33})(1 - d_{22})(-1 - d_{33}),$$
$$\det(I_3 - T_*(\mathfrak{A}_1\mathfrak{A}_2)(\mathfrak{D}_7)_*) = (-1 - d_{22}d_{33})(-1 - d_{22})(1 - d_{33}).$$

Now, Lemma 6.9 implies that the Anosov theorem holds for this case, which proves our statement concerning Family 8.

To summarize, we may conclude that the Anosov theorem holds for all, except one, families of infra-nilmanifolds in dimension 3. This result depends as before on the respective holonomy groups, but also on the structure of the universal covering Lie group which results in extra restrictions on the suitable linear parts.

The influence of the structure of the universal covering Lie group G is maximal if G is a filiform Lie group. An n-dimensional Lie group is called filiform if it is $(n-2)$-step nilpotent. Well for such a G we have a lot of structure on the affine endomorphisms and so also on the matrices we derive from them. Indeed these are $(n \times n)$-matrices where the first $n-1$ columns are completely determined by the last column and the structure of G. Obviously, one may expect that this also influences the signs of the determinants of Theorem 2.5. This leads to the following questions.

Question 8.2. *What can we say about the validity of the Anosov theorem for (orientable) infra-nilmanifolds with filiform universal covering Lie group? Or more generally: can we capture the relationship between the validity of the Anosov theorem and the structure of the universal covering Lie group?*

8.3 The 4-dimensional, 2-step infra-nilmanifolds

From dimension 4 onwards the classification is slightly more complex, since there are families which are subdivided into a finite number of subfamilies depending on the action of the generators with non-trivial linear part on the generator d (i.e. on the commutator subgroup) of the lattice N.

In order to be able to present our results, we need to agree on how we refer to a specific subfamily of a given family, say F (here F denotes a natural number as in the 3-dimensional case). An almost-Bieberbach group of Family F will be generated by a, b, c, d together with at most 3 generators with non-trivial linear part denoted by α, β and γ. If a family F is subdivided into several subfamilies, then these subfamilies are determined by the action of α, β and γ on d. Since such a generator acts or trivially on d or by inversion on d, we will denote a subfamily by

F $(\pm 1, \pm 1, \ldots)$ (where there are as many ± 1's as there are generators with non-trivial linear part), where the first ± 1 stands for the action of α on d (where of course 1 is used to denote the trivial action and -1 to indicate the non-trivial action). The second ± 1, if needed, is used to denote the action of β and analogously, if necessary the third one is used for γ. If the action on d is trivial for all generators $\alpha, \ldots$ we will omit this notation. Again we refer to [12] for the necessary details.

Just like we treated the flat manifolds in Chapter 7, we will also divide the infra-nilmanifolds into two groups: the ones with an abelian holonomy group and the ones with a non-abelian holonomy group. Of course, similar arguments as the ones used in Chapter 7 can be used to investigate both groups. Moreover, we can already mention that the investigation of the second group is very interesting since we will obtain that the Anosov theorem holds for all orientable infra-nilmanifolds of this group.

8.3.1 Abelian holonomy group

The majority of the infra-nilmanifolds in this section are covered by Theorem 5.4 (the infra-nilmanifolds with holonomy group $\mathbb{Z}_3, \mathbb{Z}_4$ and $\mathbb{Z}_6$) and by Proposition 3.7 (the non-orientable infra-nilmanifolds). The only two classes, which need to be examined are the infra-nilmanifolds with holonomy group $\mathbb{Z}_2$ and $\mathbb{Z}_2 \oplus \mathbb{Z}_2$. The calculations that need to be done here, are completely similar to the ones we did in the 3-dimensional case and for brevity we omit them. The table for the nilmanifolds is completely similar as the one in dimension 3.

Family	k	1		Non-orientable
		Orientable		
		An.th. holds	An.th. holds not	An.th. holds not
1	/	1	0	0

For the infra-nilmanifolds with $\mathbb{Z}_2$ as holonomy group we obtain the following table.

		$\mathbb{Z}_2$		
Family	k	Orientable		Non-orientable
		An.th. holds	An.th. holds not	An.th. holds not
2	$k \equiv 0 \bmod 2$	0	0	1
3	$k \equiv 0 \bmod 2$	0	1	0
4	/	0	1	0
	$k \equiv 0 \bmod 2$	0	1	0
4 (-1)	/	0	0	1
5	/	0	1	0
6	/	0	0	1
7	/	0	0	2
7 (-1)	/	2	1	0
	$k \equiv 0 \bmod 2$	0	1	0
8	/	0	0	1
9	/	0	0	1
9 (-1)	/	1	1	0

To be precise for the Families $7(-1)$ and $9(-1)$, we were able to construct a counter example for the Bieberbach groups labelled $(k, 0, 0, 0)$.

We want to focus on two types of infra-nilmanifolds of Family $7(-1)$: the ones belonging to the almost-Bieberbach group labelled $(k, 0, 0, 0)$ and $(0, k, 0, 0)$. For both of these almost-Bieberbach groups the generator with non-trivial linear part is given by $(c^{\frac{1}{2}}, \mathfrak{A})$ with $T_*(\mathfrak{A}) =$

$$\begin{pmatrix} -1 & 0 & 0 & 0 \\ 0 & 1 & 0 & 0 \\ 0 & 0 & -1 & 0 \\ 0 & 0 & 0 & 1 \end{pmatrix}.$$ This implies that at first sight they admit the same

suitable affine endomorphisms $(\delta, \mathfrak{D})$ with

$$\mathfrak{D}_* = \begin{pmatrix} d_{11} & 0 & d_{13} & 0 \\ 0 & d_{22} & 0 & d_{24} \\ 0 & 0 & d_{33} & 0 \\ 0 & d_{42} & 0 & d_{44} \end{pmatrix}$$

such that all entries are integers; $d_{24} \in 2\mathbb{Z}$ and $d_{44} \in 1 + 2\mathbb{Z}$.(As before we do not need to specify δ and we may assume that $\mathfrak{A}\mathfrak{D} = \mathfrak{D}\mathfrak{A}$). But the extra restrictions (coming from the nilpotent covering group) on the first column of the linear part $\mathfrak{D}_*$ differ. Since for the group labelled $(k, 0, 0, 0)$ we have that $[b, a] = d^k, [c, a] = 1, [c, b] = 1$ and so

$d_{11} = d_{22}d_{33}$. If we then take $\delta = 1$ and $\mathfrak{D}_* = \begin{pmatrix} 0 & 0 & 0 & 0 \\ 0 & 0 & 0 & 0 \\ 0 & 0 & 3 & 0 \\ 0 & 0 & 0 & 3 \end{pmatrix}$, we have that

the Anosov relation does not hold for the map induced by this affine endomorphism $(\delta, \mathfrak{D})$ since $\det(I_4 - \mathfrak{D}_*) = 4$ and $\det(I_4 - T_*(\mathfrak{A})\mathfrak{D}_*) = -8$.

On the other hand for the group labelled $(0, k, 0, 0)$, we have that $[b, a] = 1, [c, a] = 1, [c, b] = d^{2k}$ and so $d_{11} = d_{33}d_{44}$. If we then again calculate the determinants of Theorem 2.5, we obtain that $\det(I_4 - \mathfrak{D}_*) = (1 - d_{33}d_{44})(1 - d_{33})(1 - d_{22})(1 - d_{44})$ and $\det(I_4 - T_*(\mathfrak{A})\mathfrak{D}_*) = (1 + d_{33}d_{44})(1 + d_{33})(1 - d_{22})(1 - d_{44})$. Since d_{44} can not be zero and d_{33} is an integer, one can as before verify that these determinants have the same sign. So Theorem 2.5 implies that the Anosov theorem holds for the corresponding infra-nilmanifold.

Family $7(-1)$ clearly demonstrates that the answer to Question 8.2 will in general be far from trivial since the relationship can be very delicate.

Finally, we also want to briefly present the other counter examples by giving for each family the translational part δ and the $\mathfrak{D}_*$ of an affine endomorphism $(\delta, \mathfrak{D})$ which induces a map on the respective infra-nilmanifold that not satisfies the Anosov relation. We only focus on the orientable infra-nilmanifolds and for both cases of the Families 4 and 7 we can respectively use the same counter example.

Family	δ	$\mathfrak{D}_*$	Family	δ	$\mathfrak{D}_*$
3	1	$\begin{pmatrix} -1 & 0 & 0 & 0 \\ 0 & 1 & 0 & 1 \\ 0 & 0 & 0 & 0 \\ 0 & 2 & 0 & 1 \end{pmatrix}$	4	1	$\begin{pmatrix} -1 & 0 & 0 & 0 \\ 0 & 1 & 0 & 1 \\ 0 & 0 & 3 & 0 \\ 0 & 2 & 0 & 1 \end{pmatrix}$
5	1	$\begin{pmatrix} -1 & 0 & 0 & 0 \\ 0 & 0 & 1 & 2 \\ 0 & 1 & 0 & 2 \\ 0 & 1 & 1 & 3 \end{pmatrix}$	7 (-1)	1	$\begin{pmatrix} 0 & 0 & 0 & 0 \\ 0 & 0 & 0 & 0 \\ 0 & 0 & 3 & 0 \\ 0 & 0 & 0 & 3 \end{pmatrix}$
9 (-1)	1	$\begin{pmatrix} 0 & 0 & 0 & 0 \\ 0 & 1 & -1 & 0 \\ 0 & -1 & 1 & 0 \\ 0 & 0 & 0 & 3 \end{pmatrix}$			

For the infra-nilmanifolds with $\mathbb{Z}_2 \oplus \mathbb{Z}_2$ as holonomy group we obtain the following table.

		$\mathbb{Z}_2 \oplus \mathbb{Z}_2$		
Family	k	Orientable		Non-orientable
		An.th. holds	An.th. holds not	An.th. holds not
11	$k \equiv 0 \bmod 2$	0	0	1
13	$k \equiv 0 \bmod 2$	0	0	1
14	$k \equiv 0 \bmod 2$	0	0	2
	$k \not\equiv 0 \bmod 2$	0	0	1
14 (-1,1)	/	0	0	1
	$k \equiv 0 \bmod 2$	0	0	1
15	$k \equiv 0 \bmod 2$	0	0	1
18 (1,-1)	$k \equiv 0 \bmod 2$	0	0	1
19 (-1,1)	/	0	0	1
26 (-1,1)	/	0	0	1
27 (1,-1)	$k \equiv 0 \bmod 2$	1	0	0
29 (-1,1)	/	0	0	1
	$k \equiv 0 \bmod 2$	0	0	1
29 (1,-1)	/	0	3	0
29 (-1,-1)	/	0	0	2
30 (1,-1)	/	0	1	0
31 (-1,1)	/	0	0	1
32 (1,-1)	$k \equiv 0 \bmod 2$	2	0	0
33 (-1,1)	/	0	0	1
33 (1,-1)	/	0	2	0
33 (1,-1)	$k \equiv 0 \bmod 2$	0	1	0
33 (-1,-1)	/	0	0	1
34 (1,-1)	/	1	0	0
36 (-1,1)	/	0	0	1
37 (1,-1)	$k \equiv 0 \bmod 2$	1	0	0
41 (1,-1)	/	2	0	0
43 (1,-1)	$k \equiv 0 \bmod 2$	1	0	0
	$k \not\equiv 0 \bmod 2$	0	1	0
45 (1,-1)	$k \equiv 0 \bmod 2$	2	0	0

For these infra-nilmanifolds we want to focus to Family $43(1, -1)$ since this is the first case where the validity of the Anosov theorem depends explicitly on the value of k. The almost Bieberbach groups in this case, are generated by $a, b, c, d, \alpha, \beta$. The α, β are respectively equivalent to $(c^{\frac{1}{2}}, \mathfrak{A}_1), (b^{\frac{1}{2}}, \mathfrak{A}_2)$ with

$$T_*(\mathfrak{A}_1) = \begin{pmatrix} 1 & 2 & 0 & 2 \\ 0 & 0 & 1 & 0 \\ 0 & 1 & 0 & 0 \\ 0 & -1 & -1 & -1 \end{pmatrix} \quad \text{and} \quad T_*(\mathfrak{A}_2) = \begin{pmatrix} -1 & 0 & 0 & 0 \\ 0 & 0 & 0 & -1 \\ 0 & 1 & 1 & 1 \\ 0 & -1 & 0 & 0 \end{pmatrix}$$

The role of k is determined by $[b, a] = d^k, [c, a] = d^k, [c, b] = d^{-k}$.

Similar calculations as in Section 8.2 lead to the fact that there are only two types of affine endomorphisms which can be used to construct a counter example. The other types do not exist, because of the restrictions on the translational parts, or lead to determinants which all have the same sign. The two types are $(\delta_5, \mathfrak{D}_5)$ with

$$(\mathfrak{D}_5)_* = \begin{pmatrix} 0 & d_{12} & d_{13} & -d_{12} + d_{13} \\ 0 & d_{22} & d_{23} & -d_{22} + d_{23} \\ 0 & d_{22} & 2d_{22} - d_{23} & d_{22} - d_{23} \\ 0 & -d_{22} & d_{12} - d_{13} - d_{23} & d_{12} - d_{13} + d_{22} - d_{23} \end{pmatrix}$$

(which results from $\mathfrak{A}_1 \mathfrak{D}_5 = \mathfrak{D}_5 \mathfrak{A}_1; \mathfrak{D}_5 = \mathfrak{D}_5 \mathfrak{A}_2$) and $(\delta_6, \mathfrak{D}_6)$ with

$$(\mathfrak{D}_6)_* = \begin{pmatrix} 0 & d_{12} & d_{13} & d_{12} - d_{13} \\ 0 & d_{22} & d_{23} & d_{22} - d_{23} \\ 0 & -d_{22} + 2d_{23} & d_{23} & -d_{22} + d_{23} \\ 0 & -d_{12} + d_{13} - d_{22} & -d_{23} & -d_{12} + d_{13} - d_{22} + d_{23} \end{pmatrix}$$

(which results from $\mathfrak{A}_1 \mathfrak{D}_6 = \mathfrak{D}_6 \mathfrak{A}_1; \mathfrak{A}_1 \mathfrak{D}_6 = \mathfrak{D}_6 \mathfrak{A}_2$). The calculations and conclusions are the same for both cases, so we proceed with the first one. Rewriting the restrictions on the translational parts and using $\delta = (d_1, d_2, d_3, d_4)^t, x = \frac{-d_{22} + d_{23}}{2} + d_1 - d_2, y = \frac{d_{12} - d_{13}}{2} + 2d_1 + 2d_3$, we obtain

$$\delta \mathfrak{D}(c^{\frac{1}{2}}) \mathfrak{A}(\delta^{-1}) c^{\frac{-1}{2}} = a^x b^{-x} c^{\frac{-1}{2} + x + y} d^{-y - \frac{k}{2}x(3x - 2y)}.$$

This can only be an element of N if all the exponents are integers. So x and $\frac{-1}{2} + y$ must be integers and therefore we have that $\frac{k}{2}x(3x - 2y)$ must be an element of $\frac{1}{2} + \mathbb{Z}$. Now $x(3x - 2y)$ is also an integer because of the restrictions on x, y. Therefore if k is even, the exponent of d can never be an integer and therefore we can not construct a counter example. On the other hand, if k is odd and we carefully choose x

and y, then we can construct a counter example. For instance the map induced by $(\delta, \mathfrak{D})$ with

$$\delta = \begin{pmatrix} 0 \\ 0 \\ \frac{1}{4} \\ 0 \end{pmatrix} \quad \text{and} \quad \mathfrak{D}_* = \begin{pmatrix} 0 & 9 & 1 & -8 \\ 0 & 0 & 2 & 2 \\ 0 & 0 & -2 & -2 \\ 0 & 0 & 6 & 6 \end{pmatrix}$$

does not satisfy the Anosov relation since $\det(I_4 - \mathfrak{D}_*) = -3$ and $\det(I_4 - \mathfrak{A}_*\mathfrak{D}_*) = 5$.

Again we end with briefly describing all counter examples. Note that for the three cases of Family $29(1,-1)$ we can use the same counter example. The first entry of the second line corresponds with the almost-Bieberbach group labelled $(k, 0, 0, 0, 0)$ and the second one with the one labelled $(k, 0, 1, 0, 0)$. In the last line we consider the case $k \equiv 0 \bmod 2$.

Family	δ	$\mathfrak{D}_*$	Family	δ	$\mathfrak{D}_*$
29 (1,-1)	1	$\begin{pmatrix} 0&0&0&0 \\ 0&0&0&0 \\ 0&0&3&0 \\ 0&0&0&3 \end{pmatrix}$	30 (1,-1)	1	$\begin{pmatrix} 0&0&-3&0 \\ 0&0&0&0 \\ 0&0&3&0 \\ 0&0&0&3 \end{pmatrix}$
33(1,-1)	$a^{\frac{1}{4}}$	$\begin{pmatrix} 0&3k&0&2 \\ 0&2&0&0 \\ 0&2&0&0 \\ 0&0&0&3 \end{pmatrix}$	33(1,-1)	$a^{\frac{1}{4}}$	$\begin{pmatrix} 0&-2+3k&0&2 \\ 0&2&0&0 \\ 0&2&0&0 \\ 0&0&0&3 \end{pmatrix}$
33(1,-1)	$a^{\frac{1}{4}}$	$\begin{pmatrix} 0&-2+6k&0&2 \\ 0&4&0&0 \\ 0&4&0&0 \\ 0&0&0&3 \end{pmatrix}$			

For the remaining tables, we did not have to make explicit calculations, so we just present the results.

		$\mathbb{Z}_2 \oplus \mathbb{Z}_2 \oplus \mathbb{Z}_2$		
Family	k	Orientable		Non-orientable
		An.th. holds	An.th. holds not	An.th. holds not
56 (1,-1,1)	$k \equiv 0 \bmod 2$	0	0	1
60 (1,-1,1)	/	0	0	1
61 (-1,1,1)	$k \equiv 0 \bmod 2$	0	0	2
	$k \not\equiv 0 \bmod 2$	0	0	2
62 (-1,1,1)	$k \equiv 0 \bmod 2$	0	0	1

$\mathbb{Z}_4$				
Family	k	Orientable		Non-orientable
		An.th. holds	An.th. holds not	An.th. holds not
75	$k \equiv 0 \bmod 2$	1	0	0
	$k \equiv 0 \bmod 4$	1	0	0
76	/	1	0	0
76	$k \equiv 0 \bmod 2$	1	0	0
77	$k \equiv 0 \bmod 2$	1	0	0
79	$k \equiv 0 \bmod 2$	2	0	0
80	/	1	0	0
81	$k \equiv 0 \bmod 2$	0	0	2
	$k \equiv 0 \bmod 4$	0	0	1
82	$k \equiv 0 \bmod 2$	0	0	2
	$k \not\equiv 0 \bmod 2$	0	0	1
	$k \equiv 0 \bmod 4$	0	0	1
	$k \equiv 2 \bmod 4$	0	0	1

$\mathbb{Z}_2 \oplus \mathbb{Z}_4$				
Family	k	Orientable		Non-orientable
		An.th. holds	An.th. holds not	An.th. holds not
85	$k \equiv 0 \bmod 2$	0	0	2
86	$k \equiv 0 \bmod 2$	0	0	2
88	$k \equiv 0 \bmod 2$	0	0	2

$\mathbb{Z}_3$				
Family	k	Orientable		Non-orientable
		An.th. holds	An.th. holds not	An.th. holds not
143	$k \equiv 0 \bmod 3$	2	0	0
	$k \not\equiv 0 \bmod 3$	1	0	0
144	/	1	0	0
	$k \equiv 0 \bmod 3$	1	0	0
146	/	2	0	0

Note that we added Family 143 of [18] which is missing in [12].

$\mathbb{Z}_6$				
Family	k	Orientable		Non-orientable
		An.th. holds	An.th. holds not	An.th. holds not
147	$k \equiv 0 \bmod 6$	0	0	2
	$k \equiv 2 \bmod 6$	0	0	1
	$k \equiv 4 \bmod 6$	0	0	1
148	$k \equiv 0 \bmod 2$	0	0	2
168	$k \equiv 0 \bmod 6$	2	0	0
	$k \equiv 2 \bmod 6$	1	0	0
	$k \equiv 4 \bmod 6$	1	0	0
169	/	1	0	0
172	$k \equiv 0 \bmod 2$	1	0	0
173	$k \equiv 0 \bmod 3$	2	0	0
	$k \equiv 1 \bmod 3$	1	0	0
	$k \equiv 2 \bmod 3$	1	0	0
174	$k \equiv 0 \bmod 3$	0	0	2
	$k \not\equiv 0 \bmod 3$	0	0	1

$\mathbb{Z}_2 \oplus \mathbb{Z}_6$				
Family	k	Orientable		Non-orientable
		An.th. holds	An.th. holds not	An.th. holds not
176	$k \equiv 0 \bmod 6$	0	0	2
	$k \equiv 2 \bmod 6$	0	0	1
	$k \equiv 4 \bmod 6$	0	0	1

To summarize this section, we may say that these tables show that already many infra-nilmanifolds are covered by our results of part 2. Secondly, the relationship between the structure of the universal covering Lie group of a given infra-nilmanifold and the validity of the Anosov theorem for this manifold is very delicate. So answering Question 8.2 will not be straightforward. At least in general, but perhaps in restricted cases, such as the filiform Lie groups, the situation might become easier.

8.3.2 Non-abelian holonomy group

Part of the results in this section are again obtained by Proposition 3.7 concerning 2-step non-orientable infra-nilmanifolds. For the remaining orientable infra-nilmanifolds we find that the Anosov theorem always holds, which is rather surprising when we compare this with the analogous results in Chapter 7. The results for the infra-nilmanifolds with D_8 as holonomy group are summarized in the table below.

D_8				
Family	k	Orientable		Non-orientable
		An.th. holds	An.th. holds not	An.th. holds not
103 (1,-1)	$k \equiv 0 \bmod 2$	1	0	0
	$k \equiv 0 \bmod 4$	1	0	0
104 (1,-1)	$k \equiv 0 \bmod 2$	2	0	0
106 (1,-1)	$k \equiv 0 \bmod 2$	2	0	0
110 (1,-1)	/	2	0	0
114 (1,-1)	$k \equiv 0 \bmod 2$	0	0	2

To obtain these results we start by showing that we only have to focus on the first case of Family $103(1,-1)$, since for the other cases the suitable affine endomorphisms are very similar. This similarity leads to the fact that the determinants of Theorem 2.5 are exactly the same for all cases.

Let us first fix some notations: the generators with non-trivial linear part for the first case of Family $103(1,-1)$ are $(d^{\frac{1}{4}}, \mathfrak{A}_1)$ and $(c^{\frac{1}{2}}, \mathfrak{A}_2)$ with

$$T_*(\mathfrak{A}_1) = \begin{pmatrix} 1 & 0 & 0 & 0 \\ 0 & 0 & -1 & 0 \\ 0 & 1 & 0 & 0 \\ 0 & 0 & 0 & 1 \end{pmatrix} \quad \text{and} \quad T_*(\mathfrak{A}_2) = \begin{pmatrix} -1 & 0 & 0 & 0 \\ 0 & 1 & 0 & 0 \\ 0 & 0 & -1 & 0 \\ 0 & 0 & 0 & 1 \end{pmatrix}.$$

If we derive the matrices from the holonomy representation for the holonomy groups of the infra-nilmanifolds of Family $103(1,-1)$, Family $104(1,-1)$ and Family $106(1,-1)$, then we find that these matrices only differ in the last three elements of the first row. Of course we only compare elements which are generated in the same way in the respective holonomy groups. For instance for Family $104(1,-1)$, we have

$$T_*(\mathfrak{A}_1') = \begin{pmatrix} 1 & k & 2-k & 0 \\ 0 & 0 & -1 & 0 \\ 0 & 1 & 0 & 0 \\ 0 & 0 & 0 & 1 \end{pmatrix} \quad \text{and} \quad T_*(\mathfrak{A}_2') = \begin{pmatrix} -1 & -2 & 0 & 2 \\ 0 & 1 & 0 & 0 \\ 0 & 0 & -1 & 0 \\ 0 & 0 & 0 & 1 \end{pmatrix}$$

which clearly implies that indeed the matrices can only differ in the first row for both these cases.

As demonstrated many times before, to calculate all possibilities for the matrices $\mathfrak{D}_*$ we need these matrices of the holonomy representation. Since the respective holonomy representations are very 'similar' for each of the respective infra-nilmanifolds, the same holds for the possibilities for the $\mathfrak{D}_*$ in each case. Namely, the 'corresponding' $\mathfrak{D}_*$ also only differ in the last three elements of the first row. By corresponding we mean here, belonging to the analogous homomorphism on the respective fundamental groups. This relies completely on the fact that the last three elements of the first column of the matrices $\mathfrak{D}_*$ are all zero.

Now, if we want to verify the validity of the Anosov theorem for one of the infra-nilmanifolds of the respective families, we have to calculate the determinants of Theorem 2.5 for all suitable affine endomorphisms $(\delta, \mathfrak{D})$. Let us fix an element x of D_8, then the matrices $T_*(x)$ from the respective holonomy representations only differ in the first row and the last three elements of the first column are always zero. If we then consider for each infra-nilmanifold the corresponding $(\delta, \mathfrak{D})$, the above implies that $\det(I_n - T_*(x)\mathfrak{D}_*)$ is always the same.

At first sight, the same does not hold for Family $110(1, -1)$ but we come back to this later on.

Now, let us proceed with Family $103(1, -1)$. To calculate all suitable affine endomorphisms, we have to consider 64 possibilities. Indeed, as explained before, to construct all possible homomorphisms, we have for each of the two generators of the holonomy group 8 possibilities since D_8 has 8 elements. From the 64 possibilities, there are 56 possibilities which lead to a matrix $\mathfrak{D}_*$ of the form

$$\begin{pmatrix} 0 & 0 & 0 & d_{14} \\ 0 & 0 & 0 & d_{24} \\ 0 & 0 & 0 & d_{34} \\ 0 & 0 & 0 & d_{44} \end{pmatrix}.$$

For these 56 possibilities, the determinants in Theorem 2.5 are all equal to $(1 - d_{44})$ and so the Anosov theorem holds. The remaining 8 possibilities result in the following $\mathfrak{D}_*$

$(\mathfrak{D}_1)_* = \begin{pmatrix} (d_{22})^2 & 0 & 0 & 0 \\ 0 & d_{22} & 0 & 0 \\ 0 & 0 & d_{22} & 0 \\ 0 & 0 & 0 & d_{44} \end{pmatrix}$	$(\mathfrak{D}_2)_* = \begin{pmatrix} -(d_{22})^2 & 0 & 0 & 0 \\ 0 & d_{22} & 0 & 0 \\ 0 & 0 & -d_{22} & 0 \\ 0 & 0 & 0 & d_{44} \end{pmatrix}$
$(\mathfrak{D}_3)_* = \begin{pmatrix} (d_{22})^2 & 0 & 0 & 0 \\ 0 & 0 & -d_{22} & 0 \\ 0 & d_{22} & 0 & 0 \\ 0 & 0 & 0 & d_{44} \end{pmatrix}$	$(\mathfrak{D}_4)_* = \begin{pmatrix} -(d_{22})^2 & 0 & 0 & 0 \\ 0 & 0 & -d_{22} & 0 \\ 0 & -d_{22} & 0 & 0 \\ 0 & 0 & 0 & d_{44} \end{pmatrix}$
$(\mathfrak{D}_5)_* = \begin{pmatrix} -2(d_{22})^2 & 0 & 0 & 0 \\ 0 & d_{22} & d_{22} & 0 \\ 0 & d_{22} & -d_{22} & 0 \\ 0 & 0 & 0 & d_{44} \end{pmatrix}$	$(\mathfrak{D}_6)_* = \begin{pmatrix} 2(d_{22})^2 & 0 & 0 & 0 \\ 0 & d_{22} & -d_{22} & 0 \\ 0 & d_{22} & d_{22} & 0 \\ 0 & 0 & 0 & d_{44} \end{pmatrix}$
$(\mathfrak{D}_7)_* = \begin{pmatrix} 2(d_{22})^2 & 0 & 0 & 0 \\ 0 & d_{22} & d_{22} & 0 \\ 0 & -d_{22} & d_{22} & 0 \\ 0 & 0 & 0 & d_{44} \end{pmatrix}$	$(\mathfrak{D}_8)_* = \begin{pmatrix} -2(d_{22})^2 & 0 & 0 & 0 \\ 0 & d_{22} & -d_{22} & 0 \\ 0 & -d_{22} & -d_{22} & 0 \\ 0 & 0 & 0 & d_{44} \end{pmatrix}$

Note that the d_{ij} can be more specified by examining the restrictions on the translational parts and by considering the possibilities for the image of the lattice (under the homomorphism). We omit this since we do not need it to show the validity of the Anosov theorem.

If we calculate the determinants of Theorem 2.5 for each of these possibilities, we obtain that all determinants have the same form. Namely, if the element in the upper corner of $\mathfrak{D}_*$ is $\pm(d_{22})^2$, we have

$$(-1 + (d_{22})^2)(-1 + d_{44}) \cdot X$$

and in the other case we have

$$(-1 + 2(d_{22})^2)(-1 + d_{44}) \cdot X$$

with X always strict positive and possibly different for the different determinants. This clearly implies that for each of the respective cases that the determinants have the same sign, so we may conclude that the Anosov theorem indeed always holds.

The same conclusion holds for Family $110(1, -1)$ and this is obtained in the same way, but first we have to transform the matrices $T_*(\mathfrak{A}_1''), T_*(\mathfrak{A}_2'')$ to $PT_*(\mathfrak{A}_1'')P^{-1}, PT_*(\mathfrak{A}_2'')P^{-1}$ with

$$P = \begin{pmatrix} 2 & \frac{4+9k}{2} & \frac{-4-13k}{2} & \frac{4+9k}{2} \\ 0 & -1 & 1 & 1 \\ 0 & 1 & -1 & 1 \\ 0 & 1 & 1 & -1 \end{pmatrix}.$$

After that, we can apply exactly the same reasoning as before but we have to slightly adapt the 8 matrices $(\mathfrak{D}_i)_*$. Namely, since the structure of the lattice is different, the upper element of the $(\mathfrak{D}_i)_*$ is also different. However an analogue conclusion about the determinants holds.

It is important to note that the validity of the above results again heavily depend on the extra restrictions on the linear parts of the affine endomorphisms coming from the structure of the universal covering Lie group. Secondly, in the proof of our claims we did not use the restrictions on the translational parts.

Exactly the same arguments hold for the infra-nilmanifolds with S_3 and $\mathbb{Z}_2 \oplus S_3$ as holonomy group. The results are summarized in the tables below.

S_3				
Family	k	Orientable		Non-orientable
		An.th. holds	An.th. holds not	An.th. holds not
158 (1,-1)	$k \equiv 0 \bmod 3$	2	0	0
	$k \not\equiv 0 \bmod 3$	1	0	0
159 (1,-1)	$k \equiv 0 \bmod 3$	2	0	0
	$k \equiv 1 \bmod 3$	1	0	0
	$k \equiv 2 \bmod 3$	1	0	0
161 (1,-1)	$k \equiv 0 \bmod 3$	2	0	0
	$k \equiv 1 \bmod 3$	2	0	0
	$k \equiv 2 \bmod 3$	2	0	0

$\mathbb{Z}_2 \oplus S_3$				
Family	k	Orientable		Non-orientable
		An.th. holds	An.th. holds not	An.th. holds not
184 (1,-1)	$k \equiv 0 \bmod 6$	2	0	0
	$k \equiv 2 \bmod$	1	0	0
	$k \equiv 4 \bmod$	1	0	0

To conclude, the results obtained in this section are very intriguing since for the orientable infra-nilmanifolds in this section the Anosov theorem always holds. Since flat manifolds are a special type of infra-nilmanifolds, we know from Chapter 7 that this certainly does not hold for any infra-nilmanifold. So, in trying to generalize this, we have to focus on non-abelian Lie groups and we can reformulate Question 8.2. This restriction can be very useful since we already demonstrated that the relationship is very delicate.

Question 8.3. *Can we capture the relationship between the structure of the non-abelian universal covering Lie group and the validity of the Anosov theorem if we restrict ourselves to infra-nilmanifolds with non-abelian holonomy group?*

8.4 The 4-dimensional, 3-step infra-nilmanifolds

As in the previous section, a given family can be subdivided into sub-families according to the action on the lattice generator d, therefore, we will use an analogous notation as before. Secondly some families are also subdivided into separate classes depending on some underlying almost-crystallographic group Q which is used. When this is needed for a specific family F, we will, without any further explanation, refer to this Q by means of the labels used in [12] . For instance $F[Q = (2\ell+1, 1)](-1)$ means that in family F, there is a class of almost-Bieberbach groups labelled with $Q = (2\ell + 1, 1)$ for which the action of the generator with non-trivial linear part on d is equal to d^{-1}. For more details we refer to [12].[1]

Thirdly, we can no longer apply Proposition 3.7 on the non-orientable infra-nilmanifolds. However the main result of [19] implies that the infra-nilmanifolds that we consider in this section all admit an expanding map. So Theorem 3.6 still implies that the Anosov theorem does not hold for the non-orientable manifolds. For more details we refer to [19].

For the nilmanifolds we again we have a similar table as before.

1			
Family	k	Orientable		Non-orientable
		An.th. holds	An.th. holds not	An.th. holds not
1	/	1	0	0

---

[1] To be consistent with the previous tables,we slightly changed the notation used in [12]. To be precise, we replaced m by k and for the fourth family we replaced k by ℓ.

For the infra-nilmanifolds with $\mathbb{Z}_2$ as holonomy group we obtained the following results.

$\mathbb{Z}_2$				
Family	k	Orientable		Non-orientable
		An.th. holds	An.th. holds not	An.th. holds not
2 (-1)	/	0	0	1
3[Q=(2ℓ+1,1)]	$k \equiv 0 \bmod 2$	1	0	0
3[Q=(2ℓ,1)]	/	1	0	0
3[Q=(2ℓ,0)]	$k \equiv 0 \bmod 2$	1	0	0
4[Q=(2ℓ)]	/	1	0	0
4[Q=(2ℓ)]	$k \equiv 0 \bmod 2$	2	0	0
4[Q=(2ℓ+1)]	/	1	0	0
4[Q=(2ℓ+1)]	$k \equiv 0 \bmod 2$	1	0	0
4[Q=(2ℓ+1)]	$k \equiv 0 \bmod 4$	1	0	0
4[Q=(ℓ)] (-1)	/	0	0	ℓ
5[Q=(2ℓ,0)]	/	1	0	0
5[Q=(2ℓ+1,0)]	/	1	0	0

The prove the results for the orientable manifolds we focus on the first case of Family 3. The calculations for the other cases are analogous. Applying the techniques explained in Section 8.1 we deduce from [12] that the generator with non-trivial linear part of the infra-nilmanifolds of the case under study is equivalent with $(d^{\frac{-1}{2}}, \mathfrak{A})$ such that

$$
\mathfrak{A}_* = \begin{pmatrix} 1 & 0 & 0 & 0 \\ 0 & -1 & -1 & 0 \\ 0 & 0 & 1 & 0 \\ 0 & 0 & 0 & -1 \end{pmatrix}.
$$

Since the holonomy group is $\mathbb{Z}_2$, Proposition 2.6 implies that we only have to consider affine endomorphisms $(\delta, \mathfrak{D})$ such that $\mathfrak{D}\mathfrak{A} = \mathfrak{A}\mathfrak{D}$. This means that we may assume that for the suitable affine endomorphisms $(\delta, \mathfrak{D})$ we have that

$$
\mathfrak{D}_* = \begin{pmatrix} d_{11} & 0 & d_{23} & 0 \\ d_{21} & d_{22} & d_{23} & d_{24} \\ -d_{21} & 0 & d_{22} - 2d_{23} & 0 \\ 0 & 2d_{43} & d_{43} & d_{44} \end{pmatrix}
$$

with all entries integers. This $\mathfrak{D}_*$ does not take into account the restrictions coming from the structure of the universal covering group. Since the first column concerns the image of D and the second the image of C, we have that $d_{21} = d_{43} = 0$. Moreover d_{11}, d_{22} are completely determined by the fact that $[c, b] = d^k$ and $[b, a] = c^{2\ell+1}d^{k\ell}$ which are part of the relations we find in [12]. Indeed these relations implies that

$$k\mathfrak{D}_*(D) = [\mathfrak{D}_*(C), \mathfrak{D}_*(B)]$$
$$\Leftrightarrow kd_{11}D = [d_{22}C, d_{23}D + d_{23}C + (d_{22} - 2d_{23})A + d_{43}B]$$
$$\Leftrightarrow \qquad\qquad kd_{11}D = d_{22}d_{43}[C, B]$$
$$\Leftrightarrow \qquad\qquad kd_{11}D = kd_{22}d_{43}kD$$

and so $d_{11} = d_{22}d_{43}$ since $k > 0$. Analogously we find that

$$(2\ell + 1)\mathfrak{D}_*(C) = d_{22}C$$
$$= kd_{23}d_{44}D + (2\ell + 1)(d_{22} - 2d_{23})d_{44}C.$$

which implies that $d_{23} = 0$ or $d_{44} = 0$ and $d_{22} = (d_{22} - 2d_{23})d_{44}$.

The above results for the suitable affine endomorphisms are obtained by the restrictions on the linear parts and by the restrictions coming from the structure of the universal covering group. So there is still one type of restrictions left, namely these on the translational parts. For the first case of Family 3, we have $(\delta, \mathfrak{D})(d^{\frac{-1}{2}}, \mathfrak{A}) = (nd^{\frac{-1}{2}}, \mathfrak{A})(\delta, \mathfrak{D})$ with $n \in N$, so the restrictions on the translational parts are

$$\delta\mathfrak{D}(d^{\frac{-1}{2}})\mathfrak{A}(\delta^{-1})d^{\frac{1}{2}} \in N$$

By similar calculations as before this leads to the fact that d_{11} must be an odd integer and so d_{22} and d_{44} are non-zero integers.

To check the validity of the Anosov theorem, we use once more Theorem 2.5 and we have to determine the signs of the following two determinants.

$$\det(I_4 - \mathfrak{D}_*) = (1 - d_{22}d_{44})(1 - d_{22})(1 - d_{22} + 2d_{23})(1 - d_{44})$$
$$\det(I_4 - \mathfrak{A}_*\mathfrak{D}_*) = (1 - d_{22}d_{44})(1 + d_{22})(1 - d_{22} + 2d_{23})(1 + d_{44})$$

Because of the restrictions on the d_{22} and d_{44}, we easily obtain that the determinants have the same sign which implies that the Anosov theorem holds.

The infra-nilmanifolds with $\mathbb{Z}_2 \oplus \mathbb{Z}_2$ as holonomy group and 3-step nilpotent universal covering group are all non-orientable and so we did not have to make any calculations.

$\mathbb{Z}_2 \oplus \mathbb{Z}_2$				
Family	k	Orientable		Non-orientable
		An.th. holds	An.th. holds not	An.th. holds not
7[Q=(2ℓ)] (-1,1)	$k \equiv 0 \bmod 2$	0	0	1
7[Q=($2\ell+1$)] (-1,1)	/	0	0	1
8 (-1,1)	/	0	0	1

To conclude this section, we can say that this section further motivates to investigate the validity of the Anosov theorem for infra-nilmanifolds with filiform universal covering group.

References

1. Anosov, D. *The Nielsen numbers of maps of nil-manifolds.* Uspekhi. Mat. Nauk, 1985, 40, pp. 133–134. English transl.: Russian Math. Surveys, 40 (no. 4), 1985, pp. 149–150.
2. Brooks, R., Brown, R. F., Pak, J., and Taylor, D. *The Nielsen number of maps on tori.* Proc. Amer. Math. Soc., 1975, 52, pp. 398–400.
3. Brown, H., Bülow, R., Neubüser, J., Wondratscheck, H., and Zassenhaus, H. *Crystallographic groups of four–dimensional Space.* Wiley New York, 1978.
4. Brown, K. S. *Cohomology of groups.*, volume 87 of *Grad. Texts in Math.* Springer-Verlag New York Inc., 1982.
5. Brown, R. F. *The Lefschetz fixed point theorem.* Scott, Foresman and Company, 1971.
6. Brown, R. F. *Nielsen Fixed point theory on manifolds.* Banach Cent. Publ., 1999, 49, pp. 19–27.
7. *CARAT – Crystallographic AlgoRithms And Tables, Version 1.2.* Lehrstuhl B fur Mathematik, Aachen, 2001. (http://wwwb.math.rwth-aachen.de/carat/).
8. Charlap, L. S. *Bieberbach Groups and Flat Manifolds.* Universitext. Springer–Verlag, New York Inc., 1986.
9. Cid, C. and Schulz, T. *Computation of five- and six-dimensional Bieberbach groups.* Exp. Math., 2001, 10, 1, pp. 109–115.
10. Cobb, P. V. *Manifolds with holonomy group $\mathbb{Z}_2 \oplus \mathbb{Z}_2$ and first betti number zero.* J. Differential Geom., 1975, 10, 221–224.
11. Dekimpe, K. *The construction of affine structures on virtually nilpotent groups.* Manuscripta Math., 1995, 87 pp. 71–88.
12. Dekimpe, K. *Almost-Bieberbach Groups: Affine and Polynomial Structures*, volume 1639 of *Lect. Notes in Math.* Springer–Verlag, 1996.
13. Dekimpe, K., De Rock, B., and Malfait, W. *The Anosov theorem for flat generalized Hantzsche-Wendt manifolds.* J. Geometry and physics, 2004, 52 pp. 174–185.
14. Dekimpe, K., De Rock, B., and Malfait, W. *The Nielsen numbers of Anosov diffeomorphisms on flat Riemannian manifolds.* Forum Mathematicum, 2005, 17 pp. 325–341.
15. Dekimpe, K., De Rock, B., and Malfait, W. *The Anosov relation for Nielsen numbers of maps of infra-nilmanifolds.* Monatschefte für Mathematik, 2007, 150(1) pp. 1–10.

16. Dekimpe, K., De Rock, B., and Malfait, W. *The Anosov theorem for infra-nilmanifolds with cyclic holonomy group.* Pacific Journal of Mathematics, 2007, 52 (1) pp. 137–160.

17. Dekimpe, K., De Rock, B., and Pouseele, H. *The Anosov theorem for infra-nilmanifolds with an odd order abelian holonomy group.* Fixed Point Theory and Applications, 2006, pages pp. 1–12. article ID 63939.

18. Dekimpe, K. and Eick, B. *Computational aspects of Group extensions and their applications to topology.* Experimental Math., 2002, 11 pp. 183–200.

19. Dekimpe, K. and Lee, K. B. *Expanding maps on infra-nilmanifolds of homogeneous type.* Trans. Amer. Math. Soc., 2003, 355 (3), pp. 1067–1077.

20. Dixmier, J. and Lister, W. *Derivations of nilpotent Lie algebras.* Proc. Am. Math. Soc., 1957, 8 155–158.

21. Dugundji, J. and Granas, A. *Fixed point theory.* Springer monographs in mathematics. Springer-Verlag, New York, 2003.

22. Epstein, D. and Shub, M. *Expanding endomorphisms of flat manifolds.* Topology, 1968, 7 pp. 139–141.

23. Franks, J. *Anosov diffeomorphisms.* Global Analysis: Proceedings of the Symposia in Pure Mathematics, 1970, 14, pp. 61–93.

24. Goze, M. and Khakimdjanov, Y. *Nilpotent Lie Algebras.* Mathematics and Its Applications. Kluwer Academic Publishers, 1996.

25. Gromov, M. *Groups of polynomial growth and expanding maps.* Institut des Hautes Études Scientifiques, 1981, 53, pp. 53–73.

26. Hantzsche, W. and Wendt, H. *Driedimensionale euklidische Raumformen.* Math. Ann., 1935, 110, pp. 593–611.

27. Heath, P. R. and Keppelmann, E. C. *Model solvmanifolds for Lefschetz and Nielsen theories.* Quaest. Math., 2002, 25 4, pp. 483–501.

28. Jezierski, J., Kędra, J., and Marzantowicz. *Homotopy minimal periods for NR-solvmanifold maps.* Topology and applications, 2004, no.1-3, 29–49.

29. Jiang, B. *Nielsen Fixed Point Theory,* volume 14 of *Contemporary Math.* American Mathematical Society, 1983.

30. Jiang, B. *Commutativity and Wecken properties for fixed points of surfaces and 3-manifolds.* Topology Applications, 1993, 53 53, pp. 221–228.

31. Keppelmann, E. C. and McCord, C. K. *The Anosov theorem for exponential solvmanifolds.* Pacific J. Math., 1995, 170, No. 1, pp. 143–159.

32. Kiang, T.-h. *The Theory of Fixed Point Classes.* Springer-Verlag, 1989.

33. Kim, J. B., Seung Won; Lee and Lee, K. B. *Averaging formula for Nielsen numbers.* Nagoya Math. J., 2005, 178,.

34. Kwasik, S. and Lee, K. B. *The Nielsen numbers of homotopically periodic maps of infra-nilmanifolds.* J. London Math. Soc. (2), 1988, 38, pp. 544–554.

35. Lauret, J. *Examples of Anosov diffeomorphisms.* J. Algebra, 2003, 262 1, 201–209.

36. Lee, H. and Lee, K. B. *Expanding maps on 2-step infra-nilmanifolds.* Topology Appl., 2002, 117 (1), pp. 45–58.

37. Lee, J. B. and Lee, K. B. *Lefschetz numbers for continuous maps and periods for expanding maps on infra-nilmanifolds.* Preprint.

38. Lee, K. B. *Maps of infra-nilmanifolds.* Pacific J. Math., 1995, 168, No. 1, pp. 157–166.

39. Lee, K. B. and Raymond, F. *Rigidity of almost crystallographic groups.* Contemporary Math. A. M. S., 1985, 44, pp. 73–78.

40. Mal'cev, A. I. *On a class of homogeneous spaces.* Translations A.M.S., 1951, 39, pp. 1–33.

41. Malfait, W. *Flat manifolds with prescribed first Betti number admitting Anosov diffeomorphisms.* Monatsh. Math., 2001, 133 157–162.

42. Malfait, W. *The Nielsen numbers of virtually unipotent maps on infranilmanifolds.* Forum Math., 2001, 13, pp. 227–237.

43. Manning, A. *There are no new Anosov diffeomorphisms on tori.* Amer. J. Math., 1974, 96 (3), pp. 422–429.

44. McCord, C. K. *Computing Nielsen numbers.* Cont. Math. A.M.S., 1993, 152, pp. 249–267.

45. McCord, C. K. *Nielsen numbers of homotopically periodic maps on infrasolvmanifolds.* Proc. Amer. Math. Soc., 1994, 120, No. 1, pp. 311–315.

46. Miatello, R. and Rossetti, J. *Isospectral Hantzsche-Wendt manifolds.* J. Reine Angew. Math., 1999, 515 pp. 1–23.

47. Miatello, R. and Rossetti, J. *Comparison of Twisted P-Form Spectra for flat manifolds with Diagonal Holonomy.* Ann. Global Ana. Geom., 2002, 21 4, pp. 341–376.

48. Norton-Odenthal, B. *A product formula for the generalized Lefschetz number.* Ph. D. Thesis, University of Wisconsin, Madison, 1991.

49. Porteous, H. L. *Anosov diffeomorphisms of flat manifolds.* Topology, 1972, 11, pp. 307–315.

50. Raghunathan, M. S. *Discrete Subgroups of Lie Groups,* volume 68 of *Ergebnisse der Mathematik und ihrer Grenzgebiete.* Springer-Verlag, 1972.

51. Rossetti, J. and Szczepański, A. *Generalized Hantzsche-Wendt flat manifolds.* to appear in Revista Matem. Iberoam.

52. Warner, F. W. *Foundations of differentiable manifolds and Lie groups.* Scott, Foresman and Company, Glenview, Illinois, 1971.

53. Wecken, F. *Fixpuntklassen,III.* Math. Ann., 1942, 118, pp. 544–577.

54. Whitehead, G. *Elements of Homotopy Theory,* volume 61 of *Graduate Texts in Math.* Springer–Verlag, 1978.

55. Wong, P. *Fixed point theory for homogeneous spaces.* Amer. J. Math., 1998, 120, pp. 23–42.

Wissenschaftlicher Buchverlag bietet

kostenfreie

Publikation

von

wissenschaftlichen Arbeiten

Diplomarbeiten, Magisterarbeiten, Master und Bachelor Theses
sowie Dissertationen, Habilitationen und wissenschaftliche Monographien

ie verfügen über eine wissenschaftliche Abschlußarbeit zu aktuellen oder zeitlosen
Fragestellungen, die hohen inhaltlichen und formalen Ansprüchen genügt,
und haben **Interesse an einer honorarvergüteten Publikation**?

Dann senden Sie bitte erste Informationen über Ihre Arbeit per Email
an info@vdm-verlag.de. Unser Außenlektorat meldet sich umgehend bei Ihnen.

VDM Verlag Dr. Müller Aktiengesellschaft & Co. KG
Dudweiler Landstraße 125a
D - 66123 Saarbrücken

www.vdm-verlag.de

◎ 编辑手记

世界著名数学家 R. D. Carmichael 曾指出：

数学语言提供了表达精确思想的主要手段. 为了取得更加深刻的、影响远大的结果，必须依靠思维过程. 但若没有数学语言的支持，这种过程就无法贯彻到底. 在数学符号中好像存贮了一定量的智力，这种智力释放出来时就能产生几乎是爆炸性的威力. 这就像是强大的发动机，我们借助它竖起智力结构；如果没有这种支持，我们的能力就无法进入这种结构.

在数学语言中是有"鄙视链"存在的. 比如李群、微分拓扑、代数拓扑是位于顶端的. 本书就是一部用此工具和方法来处理问题的英文版数学专著，中文书名或可译为《内诣零流形映射的尼尔森数的阿诺索夫关系》.

　　本书的作者是布拉姆·大·罗克(Bram De Rock)，比利时人，布鲁塞尔自由大学教授，2006 年在鲁汶大学获得数学博士学位，2007年在鲁汶大学获得经济学博士学位.

　　本书作者在摘要中所述内容如下：

　　　　在本书中我们验证了内诣零流形 M 的(连续) 自映射 $f: M \to M$ 的阿诺索夫关系. 如果尼尔森数 $N(f)$ 和莱夫谢茨数 $L(f)$ 是相等的，或 $N(f) = |L(f)|$，那么映射 f 满足阿诺索夫关系. 这些数在不动点理论中找到了它们的原点，而且它们还提供了有关映射 f 的不动点的信息. 更确切地说，不动点理论的主要目标之一是计算 $MF(f)$，它是同伦于 f 的所有映射的不动点的最小数. 然而，要从它的定义计算出这个数并不容易，因此引入了下界 $N(f)$. F. 温肯和 B. 江指出，在特定条件下，内诣零流形是可以满足这个下限的，即 $MF(f) = N(f)$ (见[30]，[53]). 但是要根据它的定义计算尼尔森数是不容易的，因此我们可以尝试用另一种方法去计算它. 一种可能的方法是使用与映射 f 有关系的其他数，也就是莱夫谢茨数 $L(f)$. 1985 年，D. 阿诺索夫在[1] 中证明了下面的定理：

　　定理0.1　对任意一个内诣零流形 M 的(连续) 自映射 $f: M \to M$，我们有 $N(f) = |L(f)|$.

　　　　这是一个非常有意思的结果，因为与之前介绍的数相反，$L(f)$ 可以从它的定义计算出来. 因此对于内诣零流形的自映射 f，D. 阿诺索夫向我们展示了可以计算的 $MF(f)$. 为了不冒犯[2]的作者，我们要注意 D. 阿诺索夫的这个结果实际上是这些作者在环面结果上的概括.

　　　　由于 D. 阿诺索夫定理仅对特定类别的流形有效，因此人们自然会产生一个疑问，即该定理是否可以再次推广到其他类别的流形之中. 在这方面，D. 阿诺索夫本人已经证明了他的定理并不适用于克莱因瓶. 本书的目的是验证是否可以将阿诺索夫定理推广到内诣零流形(的某些类)中，或更笼统地说,是检验阿诺索夫关系是否适用于一个内诣零流形 M 的特殊映射 $f:M \to M$.

由于诣零流形和克莱因瓶都是内诣零流形的示例,因此对内诣零流形的推广是很自然的.此外,内诣零流形是一类经过广泛研究的流形,它除了具有其他性质外,还可以通过基于殆晶体群的代数描述来进行研究.该描述对于证明我们的结果将至关重要,因此,我们在第 1 章中回顾了内诣零流形的主要性质和定义.

在第 1 章中,我们还展示了广义内诣零流形与可解流形是不同的.注意,由于诣零流形和克莱因瓶也可以看作可解流形,因此可解流形也是一种自然推广.所以要特别注意,为可解流形发展出来的方法不能直接应用于内诣零流形,例如[27],[28],[31].

在第 2 章的预备章节中,我们回顾了上述数(与映射有关)的定义和性质.我们从一个非常笼统的描述开始,然后将这些概念应用到下面的内诣零流形映射上.事实证明,这些内诣零流形的概念已广为人知.而且 K.B.李在[38]中证明了一个很好的标准,可以验证给定的内诣零流形映射的阿诺索夫关系.这个结果对于本书来说是必不可少的,因此也将在第 2 章中进行介绍.

本书的第二部分包括了我们对阿诺索夫定理的扩展.实际上有两种可能的方式去推广阿诺索夫定理.第一种(自然的)方式是寻找流形类,而不是诣零流形,这就使该关系对已知流形的所有连续映射都成立.比如,E.吉宝曼和 C.麦科德(C. McCord)为指数可解流形建立了这一理论([31]).

在本书中,我们介绍了阿诺索夫定理所适用的几类内诣零流形.所有这些类都是通过一个内诣零流形的完整群的方式来定义的.正如我们将在第 1 章中详细解释的那样,一个内诣零流形的基本群是一个无挠的殆晶体群(称为殆比伯巴赫群),并且这些群中的一个唯一确定的有限群被称为完整群.例如,克莱因瓶的完整群为 $\mathbb{Z}_2$.

我们获得的阿诺索夫定理的一个扩展的内诣零流形的第一类,是具有奇数阶完整群的内诣零流形.该定理在第 3 章进行了验证.为了证明阿诺索夫定理对于这些流形是成立的,我们引入了一个非常有用的概念,即与映射相关的周期序列.

克莱因瓶不是流形第一类的元素,它是循环完整群的内诣零流形的一个例子,也是有限群最简单的类.在第 5 章中,我们将更仔细地研究具有循环完整群的这类内诣零流形.由于 D. 阿诺索夫的观察,我们已经知道阿诺索夫定理不是对所有类都成立.因此,我们必须引入额外的条件才能泛化阿诺索夫定理.为了找到这个额外条件,我们考虑了完整表示,该表示是一个完整群的一个准确的矩阵表示,由已知的内诣零流形(殆比伯巴赫群)来决定.为了更加准确,由于完整群是循环的,我们可以假设它通过 x_0 和使用完整表示来生成,我们可以把矩阵 $\mathbf{A}$ 与 x_0 联系起来(准确的描述请见第 1 章).那么,在这个例子中,-1 不是 $\mathbf{A}$ 的特征值,我们可以将阿诺索夫定理推广到给定的内诣零流形中.

通过这种方式,我们获得了具有循环完整群的内诣零流形的充分条件,以使阿诺索夫定理成立.请注意,克莱因瓶或任何其他以 $\mathbb{Z}_2$ 为完整群的内诣零流形不满足此条件,因为在这种情况下 -1 是关联矩阵的特征值.

对于平坦流形来说,我们可以清晰地展示这个充分条件,而且,它也是必要条件.这就意味着对于带循环完整群的平坦流形,我们得到了一个完整图.不幸的是,对于内诣零流形,情况并非如此,因此我们构建了一个以 $\mathbb{Z}_2$ 作为完整群的内诣零流形的例子(且不满足上面的条件),对于该例子阿诺索夫定理成立.事实上,我们已经在第 3 章中介绍了这个例子,之后我们会对这个例子进行讨论,在第 3 章中它是有用的.这个例子很好地说明了阿诺索夫定理在内诣零流形上的有效性比它在平坦流形上的有效性要微妙得多.

我们能够证明阿诺索夫定理成立的最后一类内诣零流形是平面可定向的广义亨茨舍—文特（Hantzsche-Wendt）流形的一类. 这是一类非常特殊的、经过充分研究的内诣零流形, 正如人们可能期望的那样, 它概括了经典的 3—维亨茨舍—文特流形. 即, 这些流形是把 $\mathbb{Z}_2^{n-1}$ 作为完整群的可定向的 n 维平坦流形. 请注意, 尽管克莱因瓶是一个以 $\mathbb{Z}_2$ 作为完整群的 2 维平坦流形, 但由于克莱因瓶是不可定向的, 因此它不是此类中的一个元素.

之所以要学习这个类, 是因为它与我们迄今为止所研究的类完全相反. 为了解释这一点, 让我们再次考虑完整表示中出现的矩阵. 因此, 对于完整群的任何元素 x_i, 我们都有一个关联的矩阵 A_i. 在我们的内诣零流形的第一类中, 奇数阶的条件可归结为以下事实: 对于任何 A_i, —1 都不是一个特征值. 对于第二类, 我们需要一个类似的条件, 但现在仅适用于循环完整群的生成元的矩阵. 在这里可能需要指出, —1 可能是完整群其他元素（非生成元）的特征值, 这会使事情变得更加复杂.

与此形成对比的是, 对于 n 维平坦可定向的广义亨茨舍—文特流形, 任何 A_i 都将 —1 作为特征值（对于许多元素, 甚至将重数 $n-1$ 作为特征值）. 尽管如此, 在第 6 章中, 我们仍然能够证明阿诺索夫定理确实适用于这类流形.

推广阿诺索夫定理的第二种途径是在某种（类型的）流形上搜索映射的类, 为此该定理成立. 例如, S. 卡西克（S. Kwasik）和 K. B. 李在 [34] 中证明了阿诺索夫定理适用于内诣零流形的同伦周期映射, 而 C. 麦科德将其扩展到内可解流形的同伦周期映射中（[45]）. 其他示例是 W. 马得胜（W. Malfait）在 [42] 中引入和研究的幂幺映射.

在本书中我们也遵循了这种方法. 在第 3 章中, 我们研究了内诣零流形的扩张映射, 并介绍了无处可扩张的内诣零流形, 正如人们所期望的那样, 它们是扩张映射的反极.

同样,通过使用与映射相关联的周期序列的概念,我们能够证明阿诺索夫定理可以推广到映射的后一类.对于扩张映射,我们获得了充要条件.准确地说,当且仅当 M 是可定向的时,阿诺索夫关系对于给定的一个内诣零流形 M 的扩张映射 $f: M \to M$ 成立.

可以使用最后的结果来试图证明阿诺索夫定理从来不适用于不可定向的内诣零流形.然而,由于有些(不可定向的)内诣零流形不允许扩张映射,因此该方法不适用于任何内诣零流形.前面提到的例子就是这种内诣零流形的例子,将在第 3 章中进行介绍.最后,要注意每一个平坦幂零内诣零流形和 2 阶的内诣零流形都接受扩张映射,如果它们是非定向的,就意味着阿诺索夫定理对于这种内诣零流形永远不成立.第 3 部分中出现的结果将非常有用.

在第 4 章中,我们研究了第三类经过充分研究的映射,即阿诺索夫微分同胚.众所周知,并非每个内诣零流形都允许这样的映射存在.到目前为止,仅对于平坦流形成立,它的阿诺索夫微分同胚的完整描述是已知的(参见[49]).因此,我们专注于平坦流形.在[41]中,已经证明了如果一个平坦流形 M 允许阿诺索夫微分同胚存在,那么他的第一个贝蒂数 $b_1(M)$ 满足下面条件中的一个:$b_1(M) = 0, 2 \leqslant b_1(M) \leqslant n - 2$,或者 $b_1(M) = n$(并且所有情况都发生).在上一个例子中 M 是一个环面,因此对于每一个 M 的连续自映射 f 都有 $N(f) = |L(f)|$.对于其他情况,我们研究了构造一个平坦流形 M 的可能性,并描述了贝蒂数,一方面,使 M 允许 $N(f) \neq |L(f)|$ 的阿诺索夫微分同胚 f 的出现,另一方面,M 也支持一个阿诺索夫微分同胚 g 满足 $N(g) = |L(g)|$.除了一些非常严格的情况,这个条件几乎是可以被满足的.即,对于原始平坦流形 $M, b_1(M) = 0$,在 6 维中,我们得出阿诺索夫定理对阿诺索夫微分同胚成立.此外,对于具有 $b_1(M) = n - 2$ 的 n 维平坦流形 M,我们得出阿诺索夫关系对于阿诺索夫微分同胚永远不成立.

在本书的第 3 部分中,我们集中讨论了低维内诣零流形,也就是 4 维以内的内诣零流形.在第 2 部分中获得的结果(或阿诺索夫的原始结果)似乎已经涵盖了许多这种内诣零流形.然而,仍然存在还没有被研究过的流形,因此,探索这些流形可为将来的研究提供灵感来源.

为了处理直到 4 维的所有内诣零流形,我们需要所有可能流形的列表.对于平坦流形,我们使用[7]的描述;对于内诣零流形(具有非阿贝尔泛覆盖群),我们使用[12]的分类.

在第 7 章中,我们考虑平坦流形,我们首先证明在 3 维中,根据第 2 部分的已知定理,我们已经有了一个完整的图像了.在 4 维中,这些定理覆盖了 74 个流形中的 53 个,其余平坦流形被分为两组:一个以 $\mathbb{Z}_2 \oplus \mathbb{Z}_2$ 作为它们的完整群,另一个是非阿贝尔完整群.第一组很有趣,因为它非常类似于我们已经研究过的内诣零流形的许多类.事实证明,该类为一些可能的问题提供了一些不错的反例,这些问题自然是从第 2 部分获得的结果中得出的.例如,我们证明了阿诺索夫定理的有效性非常微妙,不能仅从完整表示中确定.

对于第二组,如果平坦流形有 D_8 作为其完整群,则可获得类似的结果.但更有趣的是,我们发现阿诺索夫定理对于以 A_4 为完整群的(4 维)平坦流形始终成立.这引出了一个问题,即对于以 A_4 为完整群的任何平坦流形这个定理是否都成立?

对于内诣零流形,基本上需要进行相同的计算,但是存在一些复杂性.首先,平坦流形由比伯巴赫群确定,比伯巴赫群自然会以矩阵群的形式出现.对于殆比伯巴赫群,不再自动提供此矩阵表示,这使我们的计算变得复杂.其次,我们利用非阿贝尔泛覆盖群的事实对(描述该矩阵的)自同态,以及该群的仿射自同态都施加了额外的限制.这些额外的限制似乎对阿诺索夫定理的有效性产生了重要影响.例如,我们可以使用它们来证明阿诺索夫定理对于具有非阿贝尔完整群的 4 维内诣零流形一直是成立的. 当我们将其与平坦流形的相应情况进行比较时, 这是一个令人非常惊讶的结果.

这引出了一个问题, 即泛覆盖群的结构是如何影响阿诺索夫定理的有效性的.

在第 3 部分中, 我们 (几乎) 为每个比伯巴赫群提供了该特殊 (殆) 比伯巴赫群 (或内诣零流形) 的阿诺索夫关系的证明或反例. 如前所述, 对获得的结果进行分析有望为将来的研究奠定基础.

本书的版权编辑李丹女士为了方便读者阅读, 特翻译了本书的目录如下:

在国内已出版的数学专著中与本书内容最接近的似乎是江泽涵先生的《不动点类理论》.为了与其对比,我们下面摘录三段:

§1　不动点类的指数、Nielsen 数、圆周的 N 定理[①]

首先考虑圆周 S^1 的只有孤立不动点的自映射 $f:S^1 \to S^1$.

定义 1　对于 S^1 的只有孤立不动点的自映射 $f:S^1 \to S^1$,一个不动点类的全体不动点的指数和简称为此不动点类的指数.一个不动点类叫作本质的,如果它的指数不等于

---

① 摘自《不动点类理论》,江泽涵著,科学出版社.

$0, f$ 的本质的不动点类的个数叫作 f 的 Nielsen 数, 记作 $N(f)$.

引理　如果圆周 S^1 的一个自映射 f 具有圈数 n, 即

$$f \in \langle f_n \rangle$$

并且 f 只有孤立的不动点, 那么 f 的每一个不动点类的指数都是 $\mathrm{sgn}(1-n)$. 因而

$$N(f) = |1-n|$$

证明　因为 f 只有孤立的不动点, 所以它的不动点只有有限个.

设 $n < 1$. 则 f 的任一提升 $\widetilde{f_k}$ 也只有孤立的不动点, 并且只有有限个, $\sharp \Phi(\widetilde{f_k}) = \sharp p(\Phi(\widetilde{f_k}))$. 再根据定义 1, $\Phi(\widetilde{f_k})$ 的指数和等于由 $\widetilde{f_k}$ 所决定的不动点类 $p(\Phi(\widetilde{f_k}))$ 的指数.

问题归结为求 $\Phi(\widetilde{f_k})$ 的指数和. 作函数

$$\varphi(s) = s - \widetilde{f_k}(s) \tag{1}$$

由于 f 具有圈数 n, 则有

$$\widetilde{f_k}(s+1) - \widetilde{f_k}(s) = n \tag{2}$$

与

$$\varphi(s+1) - \varphi(s) = 1 - n > 0 \tag{3}$$

从而当 $s \to \pm \infty$ 时, 分别有 $\varphi(s) \to \pm \infty$, 所以式(1)分别给出 $s \gtrless \widetilde{f_k}(s)$. 由此可知

$\Phi(\widetilde{f_k})$ 的指数和 $= s$ 超过 $\widetilde{f_k}(s)$ 的次数 —

$$s \text{ 被 } \widetilde{f_k}(s) \text{ 超过的次数}$$

$$= 1$$

所以 f 的每一个不动点类的指数等于 1.

如果 $n > 1$, 相仿地可以证明 f 的每一个不动点类的指数等于 -1.

现在考虑剩下的情形: $n = 1$. 这时, 替代式(2)与(3)的

分别是

$$\widetilde{f}_k(s+1) - \widetilde{f}_k(s) = 1 \qquad (4)$$

与

$$\varphi(s+1) - \varphi(s) = 0 \qquad (5)$$

式(5)说明 $\varphi(s)$ 是周期函数,其周期是 1. 因为 f 只有孤立的不动点,所以 $\widetilde{f}_k$ 也只有孤立的不动点,于是存在一点 s_0 非 $\widetilde{f}_k$ 的不动点. 那么 $s_0 + 1$ 也非 $\widetilde{f}_k$ 的不动点. 由此结合式(5)便给出

$$\varphi(s_0 + 1) = \varphi(s_0) \neq 0 \qquad (6)$$

把 $\Phi(\widetilde{f}_k)$ 的在 φ 的一个周期 $[s_0, s_0+1]$ 上的子集记作 $\Phi'(\widetilde{f}_k)$. 则投射

$$p : \Phi'(\widetilde{f}_k) \to p(\Phi(\widetilde{f}_k))$$

是一对一的. 从而,因为式(6),有

$p(\Phi(\widetilde{f}_k))$ 的指数 $= \Phi'(\widetilde{f}_k)$ 的指数和 $=$ 在 $[s_0, s_0+1]$ 上 s 超过 $\widetilde{f}_k(s)$ 的次数 $-s$ 被 $\widetilde{f}_k(s)$ 超过的次数 $=0$

对于有非孤立不动点的自映射 f,我们也可用逼近的方法来定义 f 的 Nielsen 数.

定义 2 设 f 是 S^1 的任意一个自映射. 对于任给的充分小的数 $\varepsilon > 0$,有 f 的 ε 逼近 g,使 g 只有孤立的不动点. 这样的 g 的 Nielsen 数(根据引理,这是与 g 的选取无关的)就叫作 f 的 Nielsen 数.

如果 f 本来就只有孤立的不动点,那么可以取 f 本身作为 f 的 ε 逼近,因而定义 2 与定义 1 并没有矛盾. 把引理与定义 2 合并起来,就得到:

命题(圆周的 N 定理) 如果圆周 S^1 的自映射 f 具有圈数 n,那么

$$N(f) = |1 - n|$$

并且

$$\sharp \Phi(\langle f \rangle) = N(f)$$

§2　J 群最大时 Nielsen 数的计算

对于任意一个连通的多面体 X 及其任意一个自映射 f，我们把它们所确定的 Lefschetz 数 $L(f)$ 看作是已经算出来了，因为 $L(f)$ 是通过 f 所诱导出的、X 的同调群 $H_q(X; \mathbb{Q})$ 的自同态 f_{q*} 定义的，是能实际算出来的. 但是 X 与 f 所确定的 Nielsen 的 $N(f)$ 远不同于 $L(f)$，难以算出.

一、基本群 $\pi_1(X, x_0)$ 的自同态 $\widetilde{f}_\pi$、$\widetilde{f}_\pi$ 类、R(f) 的代数定义

我们知道，取定了 X 中的一点 x_0 作为基点后，X 的泛复迭空间 $(\widetilde{X}, p)$ 的升腾群 $\mathscr{D}(\widetilde{X}, p)$ 可以与基本群 $\pi_1(X, x_0)$ 等同起来. 取定了 $f: X \to X$ 的一个提升 $\widetilde{f}: \widetilde{X} \to \widetilde{X}$ 以后，f 的每一个提升都能唯一地表示成 $\alpha \circ \widetilde{f}$，$\alpha \in \mathscr{D} = \pi_1(X, x_0)$，这样，$f$ 的全体提升就与 $\pi_1(X, x_0)$ 建立了一一对应. 我们要探讨，这时 f 的提升对应于 $\pi_1(X, x_0)$ 中的哪些内容（结论是定理 1）.

引理 1　设 f 是连通的多面体 X 的一个自映射，并且 $\widetilde{f}$ 是 f 的一个提升. 对于任一 $\alpha \in \pi_1(X, x_0)$ 作为 $\widetilde{X}$ 的一个升腾，$\widetilde{f} \circ \alpha$ 也是 f 的一个提升，从而决定唯一的一个升腾 $\alpha'(\alpha' \in \pi_1(X, x_0))$，使得提升

$$\alpha' \circ \widetilde{f} = \widetilde{f} \circ \alpha \tag{1}$$

这单值对应 $\alpha \mapsto \alpha'$ 是 $\pi_1(X, x_0)$ 的一个自同态，叫作 X 的自映射 f 的提升 $\widetilde{f}$ 所诱导出的自同态，记作 $\widetilde{f}_\pi$.

如果提升 $\widetilde{f}$ 是由 $\widetilde{x}_0 = \langle e_0 \rangle \mapsto \widetilde{x}'_0 = \langle w'_0 \rangle$ 定义的，那么 $\widetilde{f}_\pi$ 的具体式子是

$$\widetilde{f}_\pi = w'_{0*} \circ f_\pi$$

即对于 $\langle a \rangle = \alpha$（图 1），有

$$\widetilde{f}_\pi(\alpha) = w'_{0*} \circ f_\pi(\alpha) = \langle w'_0(f \circ a)w'^{-1}_0 \rangle$$

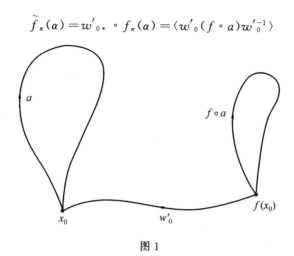

图 1

证明 设 $\alpha,\beta \in \pi_1(X,x_0)$. $\widetilde{f}_\pi$ 是自同态,因为

$$\widetilde{f}_\pi(\alpha\beta) \circ \widetilde{f} = \widetilde{f} \circ (\alpha \circ \beta) = (\widetilde{f} \circ \alpha) \circ \beta = (\widetilde{f}_\pi(\alpha) \circ \widetilde{f}) \circ \beta$$

$$= \widetilde{f}_\pi(\alpha) \circ (\widetilde{f} \circ \beta) = \widetilde{f}_\pi(\alpha) \circ (\widetilde{f}_\pi(\beta) \circ \widetilde{f})$$

$$= (\widetilde{f}_\pi(\alpha)\widetilde{f}_\pi(\beta)) \circ \widetilde{f}$$

$\widetilde{f}_\pi$ 的具体式子可如下推得

$$\langle c_x \rangle \overset{a}{\longmapsto} \langle ac_x \rangle \overset{\widetilde{f}}{\longmapsto} \langle w'_0(f \circ ac_x) \rangle = \langle w'_0(f \circ a)(f \circ c_x) \rangle$$

$$= \langle w'_0(f \circ a)w'^{-1}_0 \rangle \langle w'_0(f \circ c_x) \rangle$$

$$= (w'_{0*} \circ f_\pi)(\alpha) \circ \widetilde{f}(\langle c_x \rangle)$$

即

$$\widetilde{f} \circ \alpha = \alpha' \circ \widetilde{f} = \widetilde{f}_\pi(\alpha) \circ \widetilde{f} = (w'_{0*} \circ f_\pi)(\alpha) \circ \widetilde{f}$$

附记 1 用 $\widetilde{f}_\pi$ 的具体式子,也很容易证明 $\widetilde{f}_\pi$ 是同态.

引理 2 设 $\alpha,\alpha' \in \pi_1(X,x_0)$ 作为 $\widetilde{X}$ 的升腾,并且 $\widetilde{f}$ 是 X 的自映射 f 的一个提升. $\alpha \circ \widetilde{f}$ 与 $\alpha' \circ \widetilde{f}$ 属于 f 的同一个提升类 $\Leftrightarrow$ 存在一个 $\gamma \in \pi_1(X,x_0)$,使得在 $\pi_1(X,x_0)$ 中,有

$$\alpha' = \gamma\alpha\widetilde{f}_\pi(\gamma^{-1}) \tag{2}$$

证明 $\alpha \circ \tilde{f}$ 与 $\alpha' \circ \tilde{f}$ 属于 f 的同一个提升类 $\Leftrightarrow$ 存在一个 $\gamma \in \pi_1(X, x_0)$，使得

$$\alpha' \circ \tilde{f} = \gamma \circ (\alpha \circ \tilde{f}) \circ \gamma^{-1} = \gamma \alpha \circ (\tilde{f} \circ \gamma^{-1}) = \gamma \alpha \tilde{f}_\pi(\gamma^{-1}) \circ \tilde{f}$$

定义 1   把式 (2) 表出的 α' 与 α 之间的关系记作 $\alpha' \sim \alpha$. 容易看出这种关系 "$\sim$" 是一个等价关系，叫作由 $\tilde{f}$ 诱导出的等价关系，因而它把 $\pi_1(X, x_0)$ 划分为互不相交的等价类，叫作 $\pi_1(X, x_0)$ 的 $\tilde{f}_\pi$ 类. 含有 α 的 $\tilde{f}_\pi$ 类是

$$\{\xi \alpha \tilde{f}_\pi(\xi^{-1}) \mid \forall \xi \in \pi_1(X, x_0)\}$$

从引理 2 立刻有：

定理 1   设 f 是连通的多面体 X 的一个自映射，$\tilde{f}$ 是 f 的一个提升，并且 $\tilde{f}_\pi$ 是 $\pi_1(X, x_0)$ 的、由 $\tilde{f}$ 诱导出的自同态. $\pi_1(X, x_0)$ 的 $\tilde{f}_\pi$ 类与 f 的提升类成一一对应，因而也与 f 的不动点类成一一对应，并且 $\tilde{f}_\pi$ 类的个数由 f 唯一地决定，不依赖于所选取的提升 $\tilde{f}$.

系 1   定理 1 中 $\pi_1(X, x_0)$ 的由 f 决定的 $\tilde{f}_\pi$ 类的个数等于 Reidemeister 数 $R(f)$.

这个系可以看作 $R(f)$ 的一个代数定义.

例 1   设 f 是圆周 S^1 的具有圈数 n 的一个自映射，并且 $\tilde{f}$ 是 f 的一个提升，由 S^1 上从 x_0 到 $f(x_0)$ 的道路 w'_0 决定. $\pi_1(S^1, x_0)$ 是自由循环群，设它的母元是 γ，然后，根据引理 1 中 $\tilde{f}_\pi$ 的具体式子，有

$$\tilde{f}_\pi = w'_{0*} \circ f_\pi : \gamma \mapsto \gamma^n, \xi = \gamma^k \mapsto \gamma^{nk}, k \text{ 为整数}$$

$$\{\xi \alpha \tilde{f}_\pi(\xi^{-1}) \mid \forall \xi \in \pi_1(S^1, x_0)\} = \{\alpha \gamma^{(1-n)k} \mid k \text{ 为整数}\}$$

所以当 $n = 1$ 时，每一个群元 α 本身单独地成为一个 $\tilde{f}_\pi$ 类. 当 $n \neq 1$ 时，共有 $|1 - n|$ 个 $\tilde{f}_\pi$ 类，即取 α 依次为下列元素

$$\gamma, \gamma^2, \cdots, \gamma^{|1-n|}$$

时所得的 $\widetilde{f}_\pi$ 类. 于是按照 $n=1$ 或 $n\neq 1$, $R(f)$ 分别是无穷或 $|1-n|$.

例2 如果连通的多面体 X 是单连通的,那么 X 的任意自映射 f 的 $R(f)=1$.

例3 设连通的多面体 X 的一个自映射 f 同伦于 X 的恒同映射 id, 即 $f\simeq \mathrm{id}$. 那么

$$R(f)=R(\mathrm{id})\geqslant \sharp Z(\pi_1(X,x_0)){}^{①}\geqslant 1$$

证明 最后的"$\geqslant$"成立,因为 $Z(\pi_1(X,x_0))$ 至少含有单位元.

为了证明中间的"$\geqslant$",特别取 $\widetilde{X}$ 的恒同映射 $\widehat{\mathrm{id}}$ 作为 X 的恒同映射 id 的提升,而根据系1,$R(\mathrm{id})$ 等于 $\widehat{\mathrm{id}}_\pi$ 类的个数.但根据引理1中由式子 $\alpha'\circ\widetilde{f}=\widetilde{f}\circ\alpha$ 定义的同态 $\widetilde{f}_\pi:\alpha\mapsto \alpha'$,$\widehat{\mathrm{id}}_\pi$ 是 $\pi_1(X,x_0)$ 的恒同同构:$\alpha\mapsto \alpha'=\alpha$.

因而 α 的 $\widehat{\mathrm{id}}_\pi$ 类是

$$\{\xi\alpha\xi^{-1}\mid \xi\in\pi_1(X,x_0)\}$$

特别地,当 $\alpha\in Z(\pi_1(X,x_0))$ 时,α 本身单独地成为一个 $\widehat{\mathrm{id}}_\pi$ 类.

附记2 在例1中我们用引理1中的具体式子求得 $\widetilde{f}_\pi$ 的具体式子,而在例3中我们用引理1,由 $\alpha'\circ\widetilde{f}=\widetilde{f}\circ\alpha$ 定义的同态 $\widetilde{f}_\pi:\alpha\mapsto \alpha'$ 来说明 $\widehat{\mathrm{id}}_\pi$ 是恒同同构.这是两个不同的方法,能用一个方法求得的,也能用另一方法求得.

系1只说 $\widetilde{f}_\pi$ 类的个数 $R(f)$ 不依赖于提升 $\widetilde{f}$,现在要

---

① 群 A 的单位元素与 A 的每一元素可交换. 群 A 中的、与 A 的每一元素都可交换的那些元素,即

$$\{\alpha\in A\mid \alpha\xi\alpha^{-1}=\xi,\forall \xi\in A\}$$

形成 A 的一个子群,叫作 A 的中心,记作 $Z(A)$. 容易看出,它是一个交换群. $\sharp A$ 表示群 A 元素的个数.

补充说明自同态 $\tilde{f}_\pi$ 以及 $\tilde{f}_\pi$ 类如何依赖于 $\tilde{f}$.

定理 2 设 $\tilde{f}$ 是连通的多面体 X 的自映射 f 的一个提升. 如果 f 的另一个提升 $\tilde{f}' = \beta \circ \tilde{f}$, 这里 $\beta \in \pi_1(X, x_0)$ 作为 $\tilde{X}$ 的升腾, 那么同态 $\tilde{f}_\pi$ 与 $\tilde{f}'_\pi (= (\beta \circ \tilde{f})_\pi)$ 有下述关系

$$\tilde{f}'_\pi(\xi) = \beta \tilde{f}_\pi(\xi) \beta^{-1}, \forall \xi \in \pi_1(X, x_0)$$

证明 根据引理 1, 对于任意一个 ξ, 作为升腾, 有

$$\tilde{f}' \circ \xi = \tilde{f}'_\pi(\xi) \circ \tilde{f}' = \tilde{f}'_\pi(\xi) \circ \beta \circ \tilde{f} = (\tilde{f}'_\pi(\xi)\beta) \circ \tilde{f}$$

此外

$$\tilde{f}' \circ \xi = \beta \circ \tilde{f} \circ \xi = \beta \circ (\tilde{f} \circ \xi) =$$
$$\beta \circ (\tilde{f}_\pi(\xi) \circ \tilde{f}) = (\beta \tilde{f}_\pi(\xi)) \circ \tilde{f}$$

这就给出所求证的结论.

系 2 沿用定理 2 的假设. $\alpha, \alpha' (\alpha, \alpha' \in \pi_1(X, x_0))$ 属于同一个 $\tilde{f}'_\pi$ 类 $\Leftrightarrow \alpha\beta, \alpha'\beta$ 属于同一个 $\tilde{f}_\pi$ 类.

证明 从定义 1 与定理 2, α, α' 属于同一个 $\tilde{f}'_\pi$ 类 $\Leftrightarrow$ 存在一个 $\gamma \in \pi_1(X, x_0)$, 使得 $\alpha' = \gamma\alpha\tilde{f}'_\pi(\gamma^{-1}) = \gamma\alpha\beta\tilde{f}_\pi(\gamma^{-1})\beta^{-1} \Leftrightarrow$ 存在一个 $\gamma \in \pi_1(X, x_0)$, 使得 $\alpha'\beta = \gamma(\alpha\beta)\tilde{f}_\pi(\gamma^{-1})$.

系 3[①] 沿用定理 2 的假设. $\tilde{f}' = \beta \circ \tilde{f}$ 与 $\tilde{f}$ 属于 f 的同一个提升类, 换句话说, β 与 $\pi_1(X, x_0)$ 的单位元属于同一个 $\tilde{f}_\pi$ 类 $\Rightarrow$ 存在一个 $\gamma \in \pi_1(X, x_0)$, 使得

$$\tilde{f}'_\pi(\xi) = \gamma\tilde{f}_\pi(\gamma^{-1}\xi\gamma)\gamma^{-1} = (\gamma\tilde{f}_\pi(\gamma^{-1}))\tilde{f}_\pi(\xi)(\gamma\tilde{f}_\pi(\gamma^{-1}))^{-1}$$
$$\forall \xi \in \pi_1(X, x_0)$$

证明 由引理 2 与定义 1, 知道"$\Rightarrow$"左边两事实等价. 又 β 与单位元属于同一个 $\tilde{f}_\pi$ 类 $\Rightarrow$ 存在一个 γ, 使得

$$\beta = \gamma\tilde{f}_\pi(\gamma^{-1})$$

---

① 这个系在本节中并不需要, 但有其历史意义, 见后面的附记 3.

再由定理 2, 有

$$\tilde{f}'_\pi(\xi) = (\gamma\tilde{f}_\pi(\gamma^{-1}))\,\tilde{f}_\pi(\xi)(\gamma\tilde{f}_\pi(\gamma^{-1}))^{-1} = \gamma\tilde{f}_\pi(\gamma^{-1}\xi\gamma)\gamma^{-1}$$
$$\forall \xi \in \pi_1(X, x_0)$$

附记 3 对于 $\alpha \in \pi_1(X, x_0)$ 作为 $\widetilde{X}$ 的升腾, Nielsen 做另一升腾 (请证这是升腾) $\alpha' = \tilde{f} \circ \alpha \circ \tilde{f}^{-1}$, 得到 $\pi_1(X, x_0)$ 的一个自同构

$$I: \alpha \mapsto \alpha' = \tilde{f} \circ \alpha \circ \tilde{f}^{-1}$$

即

$$I(\alpha) \circ \tilde{f} = \tilde{f} \circ \alpha \qquad (3)$$

这表明关于任意连通的多面体 X 的任意自映射 f 的式 (1) 是 Nielsen 的式 (3) 的推广, 并且引理 1 中的 $\tilde{f}_\pi$ 是 Nielsen 的自同构 I 的推广.

如果定理 2 中的 $\tilde{f}_\pi$ 给定后, 按照 Nielsen 的术语, 把那里的全体 $\tilde{f}'_\pi$ 叫作 f 诱导的一个自同态族, 那么对于同胚 f 和 $\tilde{f}$ 就有一个自同构族.

二、$R(f)$ 的一个下界

先说明关于一般群 (不必有限, 不必交换) 的几个通用的术语与符号. 设 A, B 是群. 如果 $\varphi: A \to B$ 是从群 A 到群 B 的一个同态, A 的像 $\varphi(A)$ 是 B 的一个子群, 记作 $\mathrm{Im}\,\varphi$, 并且 B 的单位元 e 的逆像 $\varphi^{-1}(e)$ 是 A 的一个子群, 叫作 φ 的核, 记作 $\mathrm{Ker}\,\varphi$. 如果 $\mathrm{Im}\,\varphi$ 是 B 的正规子群, 那么有商群 $B/\mathrm{Im}\,\varphi$, 叫作 φ 的余核, 记作 $\mathrm{Coker}\,\varphi$.

设 f 是连通的多面体 X 的一个自映射. 对于 X 的 q 维的以整数加群为系数的同调群, f 诱导出同态 $f_{q*}: H_q(X) \to H_q(X)$; 对于 X 的基本群, f 也诱导出同态 $f_\pi: \pi_1(X, x_0) \to \pi_1(X, f(x_0))$, $\tilde{f}_\pi: \pi_1(X, x_0) \to \pi_1(X, x_0)$ 等 (见前面).

定理 3 如果 f 是连通的多面体 X 的一个自映射, 那么

$$R(f) \geqslant \# \mathrm{Coker}(1 - f_{1*})^{①} \geqslant 1$$

如果 $\pi_1(X, x_0)$ 还是交换群,那么代替左边"$\geqslant$"的是"$=$".

证明　首先,根据引理 1,$\widetilde{f}_\pi = w'_{0*} \circ f_\pi$,因此图 2 表示有交换性:

$$\pi_1(X, x_0) \xrightarrow{\widetilde{f}_\pi} \pi_1(X, x_0)$$
$$\theta \downarrow \qquad\qquad\qquad \downarrow \theta$$
$$H_1(X) \xrightarrow{f_{1*}} H_1(X)$$

图 2

其次,根据定义 1,α 所属的 $\widetilde{f}_\pi$ 类中的任意一元 α' 可以表成

$$\alpha' = \gamma \alpha \widetilde{f}_\pi(\gamma^{-1}), \gamma \in \pi_1(X, x_0)$$

所以,根据刚才证的交换性,有

$$\theta(\alpha') = \theta(\gamma \alpha \widetilde{f}_\pi(\gamma^{-1}))$$
$$= \theta(\gamma) + \theta(\alpha) + \theta(\widetilde{f}_\pi(\gamma^{-1}))$$
$$= \theta(\gamma) + \theta(\alpha) + f_{1*}(\theta(\gamma^{-1}))$$
$$= \theta(\alpha) + (1 - f_{1*})(\theta(\gamma))$$

于是,如果 α' 属于含有 α 的 $\widetilde{f}_\pi$ 类,那么存在一个 $\gamma \in \pi_1(X, x_0)$,使得

$$\theta(\alpha') - \theta(\alpha) = (1 - f_{1*})(\theta(\gamma)) \in (1 - f_{1*})(H_1(x))$$

再次,用

$$\eta : H_1(X) \to \mathrm{Coker}(1 - f_{1*}) \tag{1}$$

表示自然同态,考虑

$$\pi_1(X, x_0) \xrightarrow{\theta} H_1(X) \xrightarrow{\eta} \mathrm{Coker}(1 - f_{1*}) \tag{2}$$

这里 θ 与 η 都是满同态,因而 $\eta \circ \theta$ 也是,并且每一个 $\widetilde{f}_\pi$ 类中的全体元素的 $\eta \circ \theta$ 像都是 $\mathrm{Coker}(1 - f_{1*})$ 的同一个元

---

①　这里 $1 - f_{1*} : H_1(X) \to H_1(X)$,其中的 1 表示 $H_1(X)$ 的恒同自同构.

素. 这就给出本定理的第一个结论.

最后, 如果 $\pi_1(X, x_0)$ 是交换群, 上面证明中的交换化 θ 是一个同构. 因此 $\alpha' \sim \alpha$ 的充要条件是存在 $\gamma \in \pi_1(X, x_0)$ 使

$$\theta(\alpha') - \theta(\alpha) = (1 - f_{1*})(\theta(\gamma))$$

即

$$\eta \circ \theta(\alpha') - \eta \circ \theta(\alpha) = 0$$

因而有本定理的第二个结论.

三、$R(f) = \#\mathrm{Coker}(1 - f_{1*})$ 的条件

定义 1 所说的等价关系有一简单的运算规律:

引理 3 设 $\widetilde{f}$ 是 $f: X \to X$ 的一个提升, 关系"$\sim$"是由 $\widetilde{f}$ 所诱导的、$\pi_1(X, x_0)$ 中的等价关系. 那么, 对任意 $\alpha, \beta \in \pi_1(X, x_0)$ 有

$$\alpha\beta \sim \beta \widetilde{f}_{\pi}(\alpha)$$

特别地, 有

$$\alpha \sim \widetilde{f}_{\pi}(\alpha)$$

证明 根据定义 1, 有

$$\alpha\beta \sim (\alpha^{-1})(\alpha\beta) \widetilde{f}_{\pi}(\alpha) = \beta \widetilde{f}_{\pi}(\alpha)$$

令 $\beta = e$, 就得到 $\alpha \sim \widetilde{f}_{\pi}(\alpha)$.

本部分的主要定理是:

定理 4 设 X 是连通的多面体, $f: X \to X$ 是 X 的自映射, $\widetilde{f}$ 是 f 的提升, 关系"$\sim$"是 $\widetilde{f}$ 所诱导的、$\pi_1(X, x_0)$ 中的等价关系. 那么, 以下几个条件是等价的:

(1) $\pi_1(X, x_0)$ 中的 $\widetilde{f}_{\pi}$ 类不依赖于提升 $\widetilde{f}$ 的选取, 即 f 的任意一个提升 $\widetilde{f}'$ 诱导出的 $\widetilde{f}'_{\pi}$ 类都与已给定的提升 $\widetilde{f}$ 诱导出的 $\widetilde{f}_{\pi}$ 类相同.

(2) 对于任意 $\beta \in \pi_1(X, x_0)$, 只要 $\alpha \sim \alpha'$ 就有 $\alpha\beta \sim \alpha'\beta$.

(3) 对于任意 $\alpha,\beta,\gamma \in \pi_1(X,x_0)$ 都有 $\alpha\beta\gamma \sim \beta\alpha\gamma$.

(4) 定理 3 的证明中式(2) 表示的满同态 $\eta \circ \theta : \pi_1(X,$ $x_0) \to \mathrm{Coker}(1-f_{1*})$ 把不同的 $\widetilde{f}_\pi$ 类变成不同的元素，因而诱导出 $\pi_1(X,x_0)$ 的全体 $\widetilde{f}_\pi$ 类与 $\mathrm{Coker}(1-f_{1*})$ 的全体元素之间的一一对应.

证明 (1)$\Leftrightarrow$(2)：根据定理 2 与系 2. (2)$\Rightarrow$(3)：根据引理 3 的第二个结论，有

$$\alpha\beta \sim \widetilde{f}_\pi(\alpha\beta) = \widetilde{f}_\pi(\alpha)\widetilde{f}_\pi(\beta)$$

与

$$\widetilde{f}_\pi(\alpha) \sim \alpha$$

对于这里的后一式子，从(2) 得到

$$\widetilde{f}_\pi(\alpha)\widetilde{f}_\pi(\beta) \sim \alpha\widetilde{f}_\pi(\beta)$$

再用引理 3 的第一个结论，就得到

$$\alpha\beta \sim \beta\alpha$$

最后，再从(2)，得到

$$\alpha\beta\gamma \sim \beta\alpha\gamma$$

(3)$\Rightarrow$(4)：我们需证明，在假设(3) 成立时，如果 $\eta \circ$ $\theta(\alpha) = \eta \circ \theta(\alpha')$，那么必有 $\alpha \sim \alpha'$. 证明分成下面三步.

(i) 对任意的换位子 $[\alpha,\beta] = \alpha\beta\alpha^{-1}\beta^{-1}$ 及任意的 γ，由(3) 推出

$$[\alpha,\beta]\gamma = \alpha\beta(\alpha^{-1}\beta^{-1}\gamma) \sim \beta\alpha(\alpha^{-1}\beta^{-1}\gamma) = \gamma$$

(ii) 如果 $\theta(\gamma) = \theta(\gamma')$，那么 $\gamma \sim \gamma'$. 事实上，这时 $\gamma'\gamma^{-1} \in \mathrm{Ker}\,\theta$，$\mathrm{Ker}\,\theta$ 是 $\pi_1(X,x_0)$ 的换位子群，即它的元素都是一些换位子的乘积，因此

$$\gamma' = [\alpha_1,\beta_1][\alpha_2,\beta_2]\cdots[\alpha_k,\beta_k]\gamma$$

重复运用(i)，即得 $\gamma' \sim \gamma$.

(iii) 现在设 $\eta \circ \theta(\alpha) = \eta \circ \theta(\alpha')$. 根据 η 的定义(定理 3 证明中的式(1))，可以找到 $c \in H_1(X)$，使

$$\theta(\alpha') - \theta(\alpha) = (1-f_{1*})(c) = c - f_{1*}(c)$$

设 $c = \theta(\gamma), \gamma \in \pi_1(X, x_0)$. 根据图 2, 有

$$f_{1*}(c) = f_{1*} \circ \theta(\gamma) = \theta \circ \tilde{f}_\pi(\gamma)$$

所以

$$\theta(\alpha') = \theta(\alpha) + \theta(\gamma) - \theta(\tilde{f}_\pi(\gamma)) = \theta(\gamma \alpha \tilde{f}_\pi(\gamma^{-1}))$$

由 (ii) 即得

$$\alpha' \sim \gamma \alpha \tilde{f}_\pi(\gamma^{-1}) \sim \alpha$$

(4) $\Rightarrow$ (2): 根据定理 3, $\alpha \sim \alpha' \Rightarrow \eta \circ \theta(\alpha) = \eta \circ \theta(\alpha')$. 由于 $\eta \circ \theta$ 是同态, 所以

$$\eta \circ \theta(\alpha\beta) = \eta \circ \theta(\alpha) + \eta \circ \theta(\beta)$$
$$= \eta \circ \theta(\alpha') + \eta \circ \theta(\beta) = \eta \circ \theta(\alpha'\beta)$$

从 (4) 知 $\alpha\beta \sim \alpha'\beta$.

系 4 只要上面定理中诸条件之一成立, 就有

$$R(f) = \#\text{Coker}(1 - f_{1*})$$

证明 从其中的条件 (4) 立即得到结论.

以下讨论是保证定理 4 的条件 (3) 成立的一些充分条件.

定义 2 设 $f: X \to X$ 是连通的多面体 X 的一个自映射. 我们说 f 具有终归交换性质, 如果存在一个自然数 k, 使得

$$f_\pi^k(\alpha\beta) = f_\pi^k(\beta\alpha), \forall \alpha, \beta \in \pi_1(X, x_0) \tag{1}$$

这里 $f_\pi^k: \pi_1(X, x_0) \to \pi_1(X, f^k(x_0))$ 是自映射 $f^k = f \circ f \circ \cdots \circ f(k$ 次$): X \to X$ 所诱导的基本群同态. 式 (1) 的一个等价说法是: $\pi_1(X, x_0)$ 的 f_π^k 像是 $\pi_1(X, f^k(x_0))$ 的交换子群. 我们说 f 具有中心性质, 如果 $f_\pi: \pi_1(X, x_0) \to \pi_1(X, f(x_0))$ 的像包含在 $\pi_1(X, f(x_0))$ 的中心里.

从定义立即有: f 具有中心性质 $\Rightarrow f_\pi(\pi_1(X, x_0))$ 是交换群 $\Rightarrow f$ 具有终归交换性质; 恒同映射具有中心性质 $\Leftrightarrow$ 恒同映射具有终归交换性质 $\Leftrightarrow \pi_1(X, x_0)$ 是交换群. 如果 $\pi_1(X, x_0)$ 是交换群, 那么 X 的任一自映射具有中心性质. 这不难证明: f 具有中心性质 (或终归交换性质) 这件事不

依赖于基点 x_0 的选取；f 具有中心性质（或终归交换性质），则同伦于 f 的任一自映射也具有这个性质.

引理 4 设 $f: X \to X$ 具有终归交换性质，k 是定义 2 中所给的自然数，并且 $\tilde{f}$ 是 f 的一个提升. 那么

$$\tilde{f}_\pi^k(\alpha\beta) = \tilde{f}_\pi^k(\beta\alpha), \forall \alpha, \beta \in \pi_1(X, x_0)$$

这里的 $\tilde{f}_\pi^k = \tilde{f}_\pi \circ \tilde{f}_\pi \circ \cdots \circ \tilde{f}_\pi (k \ \text{次})$.

证明 设提升 $\tilde{f}$ 是由 $\tilde{x}_0 = \langle e_0 \rangle \mapsto \tilde{x}'_0 = \langle w'_0 \rangle$ 决定的，这里 w'_0 是 X 中从 x_0 到 $f(x_0)$ 的一条道路. 反复运用引理 1 k 次，不难知道，对于 x_0 处的圈 $c, \langle c \rangle = \gamma \in \pi_1(X, x_0)$，有

$$\begin{aligned}
\tilde{f}_\pi^k(\gamma) &= (\tilde{f}_\pi)^k(\langle c \rangle) \\
&= (w'_{0*} \circ f_\pi) \circ (w'_{0*} \circ f_\pi) \circ \cdots \circ (w'_{c*} \circ f_\pi)(\langle c \rangle) \\
&= \langle w(f^k \circ c)w^{-1} \rangle = \langle w \rangle \langle f^k \circ c \rangle \langle w^{-1} \rangle \\
&= w_* \circ (f^k)_\pi(\langle c \rangle) = w_* \circ f_\pi^k(\gamma)
\end{aligned}$$

其中 w 是首尾相接的一串道路 $w'_0, f(w'_0), \cdots, f^{k-1}(w'_0)$ 的连乘积，是从 x_0 到 $f^k(x_0)$ 的道路.

因此

$$f_\pi^k(\alpha\beta) = f_\pi^k(\beta\alpha) \Rightarrow \tilde{f}_\pi^k(\alpha\beta) = \tilde{f}_\pi^k(\beta\alpha)$$

系 5 如果 $f: X \to X$ 具有终归交换性质，特别是如果它具有中心性质，那么

$$R(f) = \# \mathrm{Coker}(1 - f_{1*})$$

证明 设 $\tilde{f}$ 是 f 的一个提升. 我们来验证定理 4 中的条件 (3). 根据引理 4，取自然数 k 使 $\tilde{f}_\pi^k(\alpha\beta) = \tilde{f}_\pi^k(\beta\alpha)$. 根据引理 3 得到

$$\begin{aligned}
\alpha\beta\gamma \sim \tilde{f}_\pi^k(\alpha\beta\gamma) &= \tilde{f}_\pi^k(\alpha\beta)\tilde{f}_\pi^k(\gamma) \\
&= \tilde{f}_\pi^k(\beta\alpha)\tilde{f}_\pi^k(\gamma) \\
&= \tilde{f}_\pi^k(\beta\alpha\gamma) \sim \beta\alpha\gamma
\end{aligned}$$

然后从系 4 得到结论.

四、J 群及有关的三个引理

设 f 是连通的多面体 X 的一个自映射. 前三部分中所讨论的是 f 的不动点类的个数 $R(f)$, 还不是 f 的本质不动点类的个数, 即 f 的 Nielsen 数 $N(f)$. 为了求得 $N(f)$, 必须弄清 f 的哪些不动点类的指数不等于 0, 这是一个十分一般的问题, 不容易直接下手.

此外, 对于圆周的自映射, 每个不动点类的指数都相等, 且不动点类的指数是同伦不变量. 由此引导我们想到, 在研究上述一般问题之前, 最好先弄清一个比较特殊的问题: 在所有可能的闭同伦 $f_t : f \simeq f$ 下, f 的哪些不动点类能互相对应, 因而有相同的指数? 譬如说, 在极端的情形, 如果 f 的所有不动点类都能在闭同伦下互相对应, 因而指数都相等, 那么 $N(f)$ 就能算出来了: 当 $L(f) \neq 0$ 时, $N(f) = R(f)$; 当 $L(f) = 0$ 时, $N(f) = 0$. 这就是后三部分的指导思想.

f 的不动点类是通过 f 的提升来定义的, 是 f 的提升的不动点集的投射. 要弄清 f 的哪些不动点类能在闭同伦下互相对应, 就该弄清 f 的哪些提升能在闭同伦下互相对应. 这就是我们要讨论的中心问题.

设 $\widetilde{f} : \widetilde{X} \to \widetilde{X}$ 是 $f : X \to X$ 的一个提升, 它是由

$$\widetilde{x}_0 = \langle e_0 \rangle \mapsto \widetilde{x}'_0 = \langle w'_0 \rangle$$

决定的(w'_0 是 x_0 至 $f(x_0)$ 的一条道路). 我们知道, 这时 f 的任一个提升都能唯一地表示成 $\alpha \circ \widetilde{f}$, 这里 $\alpha \in \mathscr{D} = \pi_1(X, x_0)$. 设 $f_t : f \simeq f$ 是从 f 到 f 的一个同伦(这种同伦叫作 f 的一个闭同伦). 设在同伦 f_t 下 f 的提升 $\widetilde{f}$ 对应于 f 的提升 $\alpha \circ \widetilde{f}$, 即同伦 $\{f_t\}$ 有一个提升 $\{\widetilde{f}_t\}$ 使 $\widetilde{f}_0 = \widetilde{f}, \widetilde{f}_1 = \alpha \circ \widetilde{f}$. 我们首先要知道这个 α 与提升 $\widetilde{f}$ 及闭同伦 $\{f_t\}$ 有什么样的关系.

由

$$\widetilde{f}_1(\widetilde{x}_0) = \langle w'_0 b \rangle$$

其中 b 表示道路 $b(t) = f_t(x_0), t \in I$,又

$$\alpha \circ \widetilde{f}(\widetilde{x}_0) = \alpha(\langle w'_0 \rangle) = \alpha \langle w'_0 \rangle$$

既然 $\widetilde{f}_1 = \alpha \circ \widetilde{f}$,就该有(图 3)

$$\langle w'_0 b \rangle = \alpha \langle w'_0 \rangle$$

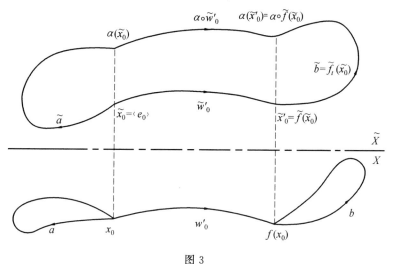

图 3

因而

$$\alpha = \langle w'_0 b w'^{-1}_0 \rangle = w'_{0*}(\langle b \rangle) \qquad (1)$$

这里的 b 是 $f(x_0)$ 处的圈,$b(t) = f_t(x_0)$,它刻画了在同伦 $\{f_t\}$ 的过程中基点 x_0 的行踪,我们把它叫作同伦 $\{f_t\}$ 的踪.以后在不致引起混淆时我们就把 $\{f_t\}$ 的踪 b 写成 $f_t(x_0)$,把 $\langle b \rangle$ 写成 $\langle f_t(x_0) \rangle$,于是式(1)可以写成

$$\alpha = w'_{0*}(\langle f_t(x_0) \rangle) \qquad (2)$$

式(2)提示我们,应该重视闭同伦的踪.这引导我们提出:

定义 3　设 f 是连通的多面体 X 的一个自映射.f 的一个闭同伦 $f_t: f \simeq f$ 决定 $f(x_0)$ 处的一个圈 $f_t(x_0), t \in I$.f 的所有闭同伦决定 $\pi_1(X, f(x_0))$ 的一个子集

$$J(f,x_0)=\{\langle f_t(x_0)\rangle\mid\forall\text{ 闭同伦 }f_t:f\simeq f\}$$

（换句话说，$\pi_1(X,f(x_0))$ 的元素 $\beta\in J(f,x_0)$ 当且仅当存在 f 的闭同伦 $f_t:f\simeq f$ 使得 $\{f_t\}$ 的踪 $f_t(x_0)$ 属于 β.）由于闭同伦的逆与积也都是闭同伦，$J(f,x_0)$ 是 $\pi_1(X,f(x_0))$ 的一个子群，叫作 X 的自映射 f 的 J 群.

例 1 圆周 S^1 的恒同映射 id 的 $J(\text{id},x_0)$ 是 $\pi_1(S^1,x_0)$.对于 $\text{id}:S^1\to S^1,x\mapsto x$，取 $h_t(x)=x\mathrm{e}^{2\pi ti}=\mathrm{e}^{2\pi(s+t)i}$，因而 $h_t:\text{id}\simeq\text{id}$ 是恒同映射的闭同伦.

此外，$h_t(x_0)=\mathrm{e}^{2\pi ti},0\leqslant t\leqslant 1$，是从 x_0 起，正向地绕 S^1 一圈，而回到 x_0 的闭道路，它代表 $\pi_1(S^1,x_0)$ 的母元 γ.根据 $J(\text{id},x_0)$ 的定义，$\gamma\in J(\text{id},x_0)\Rightarrow J(\text{id},x_0)=\pi_1(S^1,x_0)$.

例 2 圆周 S^1 的自映射 f 的 $J(f,x_0)$ 是 $\pi_1(S^1,f(x_0))$.如同例 1 的证明，取 $f_t(x)=f(x)\mathrm{e}^{2\pi ti}$，然后 $f_t:f\simeq f$，并且 $f_t(x_0)$ 代表 $\pi_1(S^1,f(x_0))$ 的母元 $\gamma,\gamma\in J(f,x_0)\Rightarrow J(f,x_0)=\pi_1(S^1,f(x_0))$.

根据式（2）及定义 3，立即得到：

引理 5 设 f 是连通的多面体 X 的一个自映射，且 $\widetilde{f}$ 是 f 的一个提升，由

$$\widetilde{x}_0=\langle e_0\rangle\mapsto\widetilde{x}'_0=\langle w'_0\rangle$$

决定.设 $\alpha\in\pi_1(X,x_0)$，且 α 也同时看成是 $\widetilde{X}$ 的一个升腾.那么，存在一个闭同伦 $f_t:f\simeq f$ 使得在同伦 $\{f_t\}$ 下 f 的提升 $\widetilde{f}$ 对应于 f 的提升 $\alpha\circ\widetilde{f}\Leftrightarrow\alpha\in w'_{0*}(J(f,x_0))$.

这样，利用 J 群，引理 5 完全回答了 f 的哪些提升能在闭同伦下互相对应的问题.下面我们考察 J 群的一些性质.

引理 6 连通的多面体的恒同映射 id 与任意自映射 f 的 J 群，即 $J(\text{id},x_0)$ 与 $J(f,x_0)$ 有下述关系：

(i) 对于 X 上从 x_0 到 $f(x_0)$ 的任意道路 w'_0，有
$$J(\text{id},x_0)\subseteq w'_{0*}(J(f,x_0))$$

(ii) $J(\text{id},x_0)=\pi_1(X,x_0)\Rightarrow J(f,x_0)=\pi_1(X,f(x_0))$.

证明 （i）设 $\alpha \in J(\mathrm{id}, x_0)$. 把引理 5 中"$\Leftarrow$"这部分应用到 $f = \mathrm{id}$ 与 $\tilde{f} = \widetilde{\mathrm{id}}$（$\tilde{X}$ 的恒同映射；作为 id 的提升，是由点道路 $w'_0 = e_0$ 所决定的）这种特殊情形，就知道存在一个闭同伦

$$h_t : \mathrm{id} \simeq \mathrm{id}$$

使得它的、由 $\widetilde{\mathrm{id}}$ 所决定的提升

$$\tilde{h}_t : \widetilde{\mathrm{id}} \simeq \alpha \circ \widetilde{\mathrm{id}}$$

现在考虑 X 的任意自映射 f 以及它的、由从 x_0 到 $f(x_0)$ 的任意道路 w'_0 所决定的提升 $\tilde{f}$. 这时，同伦

$$h_t \circ f : \mathrm{id} \circ f \simeq \mathrm{id} \circ f$$

即

$$h_t \circ f : f \simeq f$$

有提升

$$\tilde{h}_t \circ \tilde{f} : \widetilde{\mathrm{id}} \circ \tilde{f} \simeq \alpha \circ \widetilde{\mathrm{id}} \circ \tilde{f}$$

即

$$\tilde{h}_t \circ \tilde{f} : \tilde{f} \simeq \alpha \circ \tilde{f}$$

既然 f 的提升 $\tilde{f}$ 是由 w'_0 决定的，再应用引理 5 中"$\Rightarrow$"这部分，得 $\alpha \in w'_{0*}(J(f, x_0))$. 所以

$$J(\mathrm{id}, x_0) \subseteq w'_{0*}(J(f, x_0))$$

（ii）因为（i）以及 w'_{0*} 是同构即可得证.

附记 1 简单地说，（i）的证明是：从 $h_t : \mathrm{id} \simeq \mathrm{id}$ 得到同伦 $h_t \circ f : f \simeq f$.

附记 2 可以把（i）的结果直观地说成是：$J(\mathrm{id}, x_0)$ 比 $J(f, x_0)$ 小.

附记 3 （i）的证明中用的是先 $\tilde{f}$ 后 $\tilde{h}_t$ 的复合同伦. 如果先 $\tilde{h}_t$ 后 $\tilde{f}$，那么

$$\tilde{f} \circ \tilde{h}_t : \tilde{f} \circ \widetilde{\mathrm{id}} \simeq \tilde{f} \circ \alpha \circ \widetilde{\mathrm{id}} = \tilde{f} \circ \alpha = \tilde{f}_\pi(\alpha) \circ \tilde{f}$$

这就是说

$$_\pi(J(\widetilde{\mathrm{id}}, x_0)) \subseteq w'_{0*}(J(f, x_0))$$

由于 $\tilde{f}_\pi = w'_{0*} \circ f_\pi$，而且 w'_{0*} 是同构，所以

$$f_\pi(J(\mathrm{id}, x_0)) \subseteq J(f, x_0)$$

例 3　本例的目的是，对于 $X = S^1$ 这个特例，具体地说明引理 6 中 (i) 的证明的第一段. 沿用例 1 中的记号，取 $h_t(x) = x\mathrm{e}^{2\pi t i}$；$h_t : \mathrm{id} \simeq \mathrm{id}, h_t(x_0)$ 代表 $\pi_1(S^1, x_0)$ 的母元 γ.

取 id 的恒同提升 $\widetilde{\mathrm{id}} : \mathbb{R}^1 \to \mathbb{R}^1, s \longmapsto s$，我们说同伦 h_t 的由 $\widetilde{\mathrm{id}}$ 决定的提升是

$$\tilde{h}_t(s) = s + t$$

因为 p 是指数映射以及

$$p(\tilde{h}_t(s)) = \mathrm{e}^{2\pi(s+t)i} = \mathrm{e}^{2\pi s i} \cdot \mathrm{e}^{2\pi t i} = x \cdot \mathrm{e}^{2\pi t i} = h_t(x)$$

所以

$$\tilde{h}_t : \widetilde{\mathrm{id}} \simeq \gamma \circ \widetilde{\mathrm{id}}$$

例 4　本例的目的是，对于 $X = S^1$ 这个特例具体地说明引理 6 中 (i) 的证明的第二段. 如同例 2，沿用例 1 的记号. 现在设 f 是 S^1 的任一自映射. 这时候，取 $f_t(x) = f(x)\mathrm{e}^{2\pi t i}$，$f_t : f \simeq f$. 如果 $\tilde{f} : \mathbb{R}^1 \to \mathbb{R}^1$ 是 f 的任一提升，那么

$$p(\tilde{f}(s)) = f(p(s)) = f(\mathrm{e}^{2\pi s i}) = f(x)$$

我们说 $f_t(x)$ 的、由 $\tilde{f}$ 决定的提升是

$$\tilde{f}_t(s) = \tilde{f}(s) + t$$

因为

$$p(\tilde{f}_t(s)) = \mathrm{e}^{2\pi(\tilde{f}(s)+t)i} = \mathrm{e}^{2\pi \tilde{f}(s)i} \cdot \mathrm{e}^{2\pi t i} = p(\tilde{f}(s))\mathrm{e}^{2\pi t i} = f(x)\mathrm{e}^{2\pi t i}$$

所以

$$\tilde{f}_t : \tilde{f} \simeq \gamma \circ \tilde{f}$$

前文脚注中，已经提过一个群 A 的中心 $Z(A)$ 的定义，现在需要推广这个概念. 设群 B 是 A 的一个子群. 群 A 的那些元素，与 B 的每一元素可交换，即

$$\{\alpha \in A \mid \alpha \xi \alpha^{-1} = \xi, \forall \xi \in B\}$$

形成 A 的一个子群,叫作 B 在 A 中的中心化子,记作 $Z(B,A)$. 容易看出,$Z(A,A)=Z(A)$ 是交换群,而且,对于 $C \subseteq Z(A)$,$Z(C,A)=A$ 不必是交换群.

引理 7　(i)$J(f,x_0)$ 的每一元与 $f_\pi(\pi_1(X,x_0))$ 的每一元可交换,换句话说

$$J(f,x_0) \subseteq Z(f_\pi(\pi_1(X,x_0)),\pi_1(X,f(x_0)))$$

(ii)$J(\mathrm{id},x_0) \subseteq Z(\pi(X,x_0))$.

证明　设 f 的一个提升 $\tilde{f}$ 是由 $\tilde{x}_0 = \langle e_0 \rangle \mapsto \tilde{x}'_0 = \langle w'_0 \rangle$ 决定的,这里 w'_0 是 X 上从 x_0 到 $f(x_0)$ 的一条道路. 根据引理 1 中的具体式子以及 w'_{0*} 是同构,(i) 等价于下面的 (i′).

(i′)$w'_{0*}(J(f,x_0))$ 的每一元与 $\tilde{f}_\pi(\pi_1(X,x_0))$ 的每一元可交换,即

$$w'_{0*}(J(f,x_0)) \subseteq Z(\tilde{f}_\pi(\pi_1(X,x_0)),\pi_1(X,x_0))$$

我们现在来证明 (i′).

设 $\alpha \in w'_{0*}(J(f,x_0))$,我们来证 α 可与 $\tilde{f}_\pi(\pi_1(X,x_0))$ 的每一元交换. 根据引理 5,存在闭同伦 $f_t:f \simeq f$,使得 f_t 的由 $\tilde{f}$ 决定的提升

$$\tilde{f}_t:\tilde{f} \simeq \alpha \circ \tilde{f}$$

再者,对于任一 $\gamma \in \pi_1(X,x_0)$,作为 $\tilde{X}$ 的升腾,有同伦

$$
\begin{aligned}
\tilde{f}_t \circ \gamma(\text{记作 } \tilde{f}'_t):\tilde{f} \circ \gamma &\simeq (\alpha \circ \tilde{f}) \circ \gamma = \alpha \circ (\tilde{f} \circ \gamma) \\
&= \alpha \circ (\tilde{f}_\pi(\gamma) \circ \tilde{f}) \\
&= (\alpha \tilde{f}_\pi(\gamma)) \circ \tilde{f}
\end{aligned}
\tag{1}
$$

这里的等式来自引理 1. 同样地,$\tilde{f}_\pi(\gamma)$ 作为升腾,有

$$
\begin{aligned}
\tilde{f}_\pi(\gamma) \circ \tilde{f}_t(\text{记作 } \tilde{f}''_t):\tilde{f}_\pi(\gamma) \circ \tilde{f} &\simeq \tilde{f}_\pi(\gamma) \circ (\alpha \circ \tilde{f}) \\
&= (\tilde{f}_\pi(\gamma) \circ \alpha) \circ \tilde{f} \\
&= (\tilde{f}_\pi(\gamma)\alpha) \circ \tilde{f}
\end{aligned}
\tag{2}
$$

现在我们要从式(1)与(2)来证明

$$\alpha \tilde{f}_\pi(\gamma) = \tilde{f}_\pi(\gamma)\alpha \qquad (3)$$

因为 α 与 γ 分别是 $w'_{0*}(J(f,x_0))$ 与 $\pi_1(X,x_0)$ 的任意元素,式(3)就是(i′).其证明如下.

首先 $\tilde{f}'_t$ 与 $\tilde{f}''_t$ 都是 f_t 的提升,由于 γ,$\tilde{f}_\pi(\gamma)$ 作为升腾,所以

$$p(\tilde{f}'_t(\tilde{x})) = p(\tilde{f}_t(\gamma(\tilde{x}))) = f_t(p(\gamma(\tilde{x}))) = f_t(x)$$

$$p(\tilde{f}''_t(\tilde{x})) = p(\tilde{f}_\pi(\gamma)(\tilde{f}_t(\tilde{x}))) = p(\tilde{f}_t(\tilde{x})) = f_t(x)$$

其次,$\tilde{f}'_0 = \tilde{f} \circ \gamma$ 与 $\tilde{f}''_0 = \tilde{f}_\pi(\gamma) \circ \tilde{f}$ 是相同的提升(引理 1).最后根据同伦的提升由 f_0 的提升(这里 $\tilde{f}'_0 = \tilde{f}''_0$)唯一地决定,有 $\tilde{f}'_t = \tilde{f}''_t$.特别地,有 $\tilde{f}'_1 = \tilde{f}''_1$.这就给出式(3).

(ii) 当 f 是恒同映射 id 时,f_π 是 $\pi_1(X,x_0)$ 的恒同同构.从(i)与 $Z(A,A) = Z(A)$,得证.(张慧全同志得到了引理 6 与引理 7 的简单的"直接"证明.)

附记 4 引理 7 中(i)的证明已给出 $\tilde{f}_1 \circ \gamma = \tilde{f}_\pi(\gamma) \circ \tilde{f}_t$.只要注意到 $\tilde{f}_t \circ \gamma = (\tilde{f}_t)_\pi(\gamma) \circ \tilde{f}_t$,就有

$$(\tilde{f}_t)_\pi = \tilde{f}_\pi$$

例 5 $\tilde{h}_t$ 用 165 页附记 2 中所说的两个方法的后一个,容易看出:对于例 3 中 $h_t: s \longmapsto s+t$ 与例 4 中 $\tilde{f}_t: s \longmapsto f(s)+t$,有

$$(\tilde{h}_t)_\pi = (\tilde{h}_0)_\pi \quad 与 \quad (\tilde{f}_t)_\pi = \tilde{f}_\pi$$

五、J 群最大时 Nielsen 数的计算

前一部分开始时提出了一个特殊的问题:给定了连通的多面体 X 的一个自映射 f,在所有可能的闭同伦 H(即 $H: f \simeq f$)下 f 的哪些不动点类互相对应,因而它们的指数相等?同时还指出了,引理 5 可以认为是这特殊问题的完全解答.有了前一部分的准备,本部分将求出最极端(即 J

群最大,它蕴含所有不动点类的指数都相等)的,也是最简单的情形时的 $N(f)$.

定理 5   如果连通的多面体 X 的一个自映射 f 的

$$J(f,x_0)=\pi_1(X,f(x_0))$$

那么：

(i) f 具有中心性质,从而 $R(f)=\#\mathrm{Coker}(1-f_{1*})\geqslant 1$.

(ii$_a$) $L(f)\neq 0$ 时, $N(f)=R(f)\geqslant 1$,并且 f 的每一个不动点类的指数都是 $\dfrac{L(f)}{N(f)}\neq 0$.

(ii$_b$) $L(f)=0$ 时, $N(f)=0$.

证明   (i) 本定理的假设与引理 7 中 (i) 的结合给出 $\pi_1(X,f(x_0))$ 的每一元与 $f_\pi(\pi_1(X,x_0))$ 的每一元可交换,这蕴含

$$f_\pi(\pi_1(X,x_0))\subseteq Z(\pi_1(X,f(x_0)))$$

即 f 具有中心性质. 从而,根据系 5,有

$$R(f)=\#\mathrm{Coker}(1-f_{1*})\geqslant 1$$

(ii) 本定理的假设与引理 5 结合就给出:给定 f 的任意两个提升,都存在一个闭同伦 $H:f\simeq f$,使得这两个提升在 H 下互相对应,从而 f 的任意两个(从而,所有的)不动点类的指数都相等,设这个指数是 ν. 但 f 的所有不动点类指数的和就是 $L(f)$,即 $\nu\cdot R(f)=L(f)$.

(ii$_a$) $L(f)\neq 0$ 时, $\nu\neq 0$,从而每一个不动点类都是本质的,即 $N(f)=R(f)=\#\mathrm{Coker}(1-f_{1*})$,并且 $\nu=\dfrac{L(f)}{N(f)}$.

(ii$_b$) $L(f)=0$ 时, $\nu=0$,从而每一个不动点类都是非本质的,即 $N(f)=0$.

定理 6   如果连通的多面体 X 的恒同映射 id 的

$$J(\mathrm{id},x_0)=\pi_1(X,x_0)$$

那么：

(i) $\pi_1(X,x_0)$ 是交换群;

(ii) 对于 X 的任一自映射 f, $J(f,x_0)=\pi_1(X,f(x_0))$, 从而定理 5 中的结论成立.

证明 (i) 本定理的假设与引理 7 中 (ii) 的结合就是
$$\pi_1(X,x_0)\subseteq Z(\pi_1(X,x_0))$$
即 $\pi_1(X,x_0)$ 是交换群.

(ii) 从本定理的假设, 引理 6 的 (ii), 与定理 5 即得.

例 1 连通的多面体 X 的常自映射 $c:X\to x_0$, 当然, c 只有一个不动点 x_0, 从而只有一非空的不动点类. 这个不动点的指数 $\nu(c,x_0)$ 是 $+1$, 所以只有一个本质的不动点类, 即 $N(c)=1$.

当然, 应用定理 5, 也可得到这一结果. 这里值得说明, 且我们现在只说明的是, $J(c,x_0)=\pi_1(X_1,x_0)$. 设 $\alpha=\langle a\rangle\in\pi_1(X,x_0)$, 这里 a 表示 X 上在 x_0 处的一条闭道路 $a(t)$, $0\leqslant t\leqslant 1$, $a(0)=a(1)=x_0$. 作同伦
$$f_t:c\simeq c, f_t(x)=a(t)$$
根据 $J(f,x_0)$ 的定义, $\alpha\in J(c,x_0)$, 于是 $J(c,x_0)=\pi_1(X,x_0)$.

同时, 顺带指出 $f_\pi(\pi_1(X,x_0))=\{c\}$, c_{1*} 是零同态, $R(f)=\#\mathrm{Coker}(1-c_{1*})=1$.

例 2 圆周 S^1 具有圈数 $n(n\neq1)$ 的自映射 f. 从 175 页的例 2 (或例 1) 可应用定理 5 (或定理 6). 易知 $L(f)=1-n$, $N(f)=R(f)=|1-n|$ (参看 164 页例 1). 然后按照 $n<1$ 或 $n>1$, 分别有 $\nu=\dfrac{L(f)}{N(f)}=+1$ 或 -1.

例 3 $2n+1$ 维的透镜空间 $L(m;q_1,\cdots,q_n)$ 的自映射 f.

透镜空间的定义 S^{2n+1} 表示 $2n+2$ 维欧几里得空间中半径为 1 的球面. 在用 $n+1$ 个复坐标
$$z_0=r_0\mathrm{e}^{i\theta_0},\cdots,z_n=r_n\mathrm{e}^{i\theta_n}$$
时, 它的方程是
$$r_0^2+r_1^2+\cdots+r_n^2=1$$

设 $m \geqslant 2$ 是一个给定的整数，$q_1, \cdots, q_n$ 是 n 个给定的与 m 互素的整数. 考虑 S^{2n+1} 的一个旋转

$$\gamma : \begin{cases} z_0 \longmapsto \mathrm{e}^{\frac{2\pi \mathrm{i}}{m}} z_0 \\ z_j \longmapsto \mathrm{e}^{\frac{2\pi \mathrm{i} q_j}{m}} z_j, j=1, \cdots, n \end{cases}$$

作为母元，γ 生成一个 m 阶的循环群，是 S^{2n+1} 上的一个变换群，记作 π_1. π_1 的每一个元素，除去单位元，都无不动点. 把 S^{2n+1} 上的在 π_1 中任意一个旋转下相对应的那些点看作同一点，就得到我们所说的透镜空间 $L(m; q_1, \cdots, q_n)$，简记为 L，它是 $2n+1$ 维的可定向的流形. S^{2n+1} 是 L 的 m 叶的泛复迭空间，π_1 是 L 的基本群 $\pi_1(L, x_0)$，也是作为 L 的泛复迭空间的 S^{2n+1} 的升腾群.

首先，$J(\mathrm{id}, x_0) = \pi_1(L, x_0)$. 事实上，如果考虑 L 的恒同映射 id 的提升 $\widetilde{\mathrm{id}}$，S^{2n+1} 的恒同映射与下述同伦

$$\gamma_t : \begin{cases} z_0 \longmapsto \mathrm{e}^{\frac{2\pi \mathrm{i} t}{m}} z_0 \\ z_j \longmapsto \mathrm{e}^{\frac{2\pi \mathrm{i} t q_j}{m}} z_j \end{cases}$$

显然有

$$\gamma_t : \widetilde{\mathrm{id}} \simeq \gamma \circ \widetilde{\mathrm{id}}$$

根据引理 5，$\gamma \in J(\mathrm{id}, x_0)$. (读者请注意，181 页的例 1 与 175 页的例 1 和例 2 都是根据 $J(f, x_0)$ 的定义，并未用引理 5.) 从而 $J(\mathrm{id}, x_0) = \pi_1(L, x_0)$，然后可以用定理 6.

其次，应用系 4 来计算 L 的自映射 f 的 $R(f)$. 设通过 f 的一个提升 $\widetilde{f}$ 所诱导出的 $\pi_1(L)$ 的一个自同态是

$$\widetilde{f}_\pi : \gamma \longmapsto \gamma^r, 0 \leqslant r < m$$

把 $H_1(L)(\approx J_m$，整数 $\mathrm{mod}\ m$ 群) 的母元记作 θ (代替 γ 的交换化 $\theta(\gamma)$)，然后由图 2 中的交换性有 $f_{1*} : \theta \longmapsto r\theta; 1 - f_{1*} : \theta \longmapsto (1-r)\theta, x\theta \longmapsto (1-r)x\theta$，这里 x 是整数 $\mathrm{mod}\ m$. 数论中关于同余式的一条定理是，方程

$$(1-r)x \equiv 0 (\mathrm{mod}\ m)$$

恰有 $(1-r, m)$ (表示 $1-r$ 与 m 的最大公因子，是一个正整

数）个解. 于是

$$\#(1-f_{1*})H_1 = \frac{m}{(1-r,m)}$$

$$R(f) = (1-r,m)$$

再次，求 $L(f)$. 因为 $L(f) = 1 - \deg f$，所以问题归结为求 $\deg f$. 关于透镜空间，P. Olum 证明了一条定理，它包含下述事实：如果 f 诱导出

$$\tilde{f}_\pi : \gamma \mapsto \gamma^r, 0 \leqslant r < m$$

那么

$$\deg f = r^{n+1} + km$$

这里 k 是整数.

最后，应用定理 5 有：如果 $\deg f = r^{n+1} + km$ 使得 $L(f) = 1 - r^{n+1} - km \neq 0$，那么 $N(f) = (1-r,m)$，并且

$$\nu = \frac{(1 - r^{n+1} - km)}{(1-r,m)}$$

如果使得 $L(f) = 1 - r^{n+1} - km = 0$，那么 $N(f) = 0$.

六、前面两定理的应用

如果对于一个拓扑空间 X，存在一个映射

$$\mu : X \times X \to X$$

并且 X 中有一点 e，使得

$$\mu(e,x) = \mu(x,e) = x, \forall x \in X$$

那么 X 就叫作一个 H 空间，e 叫作单位点，μ 叫作乘法.

H 空间是李群以及群流形的推广.

定理 7 如果连通的多面体 X 是一个 H 空间，那么

$$J(\mathrm{id}, e) = \pi_1(X, e)$$

从而定理 6 的结论对于 X 及其任意的自映射 f 成立.

证明 e 可用来作为 X 的基点. 用 a 表示 X 上点 e 处的任一闭道路 $a(t)$，$0 \leqslant t \leqslant 1$，$a(0) = a(1) = e$. 利用乘法，定义 X 的恒同映射 id 的下述闭同伦，即

$$f_t(x) = \mu(a(t), x) : \mathrm{id} \simeq \mathrm{id}$$

明显地，$f_t(e) = a(t)$. 根据 $J(\mathrm{id}, e)$ 的定义，点 e 处的任一闭

道路 a 的 $\langle a \rangle \in J(\mathrm{id}, e)$.

定理 8   如果连通的多面体 X 是一个拓扑群对于一个道路连通的闭子群的商空间,那么

$$J(\mathrm{id}, x_0) = \pi_1(X, x_0)$$

从而定理 6 的结论对于 X 及其任意的自映射 f 成立.

如果一个拓扑空间 X 的同伦群 $\pi_n(X, x_0)$ 都是零群 $(n > 1)$,那么就说 X 是非球面性的.

引理 8   如果连通的多面体 X 是非球面性的,那么对于 X 的自映射 f,有

$$J(f, x_0) = Z(f_\pi(\pi_1(X, x_0)), \pi_1(X, f(x_0)))$$

特别地,有

$$J(\mathrm{id}, x_0) = Z(\pi_1(X, x_0))$$

证明   根据引理 7 的结论 (i),我们只需要证明

$$J(f, x_0) \supseteq Z(f_\pi(\pi_1(X, x_0)), \pi_1(X, f(x_0)))$$

从它推出下述计算 $N(f)$ 的重要定理:

定理 9   如果连通的多面体 X 是非球面性的,并且 X 的一个自映射 f 具有中心性质,那么

$$J(f, x_0) = \pi_1(X, f(x_0))$$

从而定理 5 的结论对于 X 与 f 成立.

证明 f 具有中心性质,即

$$f_\pi(\pi_1(X, x_0)) \subseteq Z(\pi_1(X, f(x_0)))$$

根据引理 7 之前紧接着的一个事实,$C \subseteq Z(A) \Rightarrow Z(C, A) = A$,我们从引理就得到 $J(f, x_0) = \pi_1(X, f(x_0))$.

§3   另一种 Nielsen 数 $N(f, H)$、根类

一、基本假设、定义与定理

设 X 是连通的多面体. 设 $(\widetilde{X}, p)$ 是 X 的一个 k(自然数)叶的正则复迭空间;换句话说,对于任意的 $x_0 \in X$,$\widetilde{x}_0 \in p^{-1}(x_0)$,$H(X, x_0) = p_\pi(\pi_1(\widetilde{X}, \widetilde{x}))$ 是 $\pi_1(X, x_0)$ 的一

个正规子群,并且 $\sharp \dfrac{\pi_1(X,x_0)}{H(X,x_0)}=k$. 如果 $\widetilde{x}_1,\widetilde{x}_2 \in p^{-1}(x_0)$, 恰存在 $\widetilde{X}$ 的一个升腾 γ,使得 $\gamma(\widetilde{x}_1)=\widetilde{x}_2$,并且升腾群 $\mathscr{D}(\widetilde{X},p) \approx \dfrac{\pi_1(X,x_0)}{H(X,x_0)}$,从而 $\sharp \mathscr{D}(\widetilde{X},p)=k$.

我们只考虑 X 的自映射 f,使得

$$f_\pi(H(X,x_0)) \subseteq H(X,f(x_0)) \tag{1}$$

如果 $f(x_0)=x'_0$,那么对于任意点 $\widetilde{x}_0 \in p^{-1}(x_0)$,$\widetilde{x'}_0 \in p^{-1}(x'_0)$,$f$ 在 $\widetilde{X}$ 上恰有一个提升 $\widetilde{f}$(即满足下述方程

$$p \circ \widetilde{f}=f \circ p \tag{2}$$

的 $\widetilde{f}:\widetilde{X} \to \widetilde{X}$),使得

$$\widetilde{f}:(\widetilde{X},\widetilde{x}_0) \to (\widetilde{X},\widetilde{x'}_0)$$

从而每一个 f 恰有 k 个不同的提升.

以上是一些基本的假设与事实.本节前两部分都设 X, $(\widetilde{X},p)$,f 满足上面的基本假设.

首先,如果 $\widetilde{f}$ 是 f 在 $\widetilde{X}$ 上的一个提升,那么

$$p(\Phi(\widetilde{f})) \subseteq \Phi(f)$$

如果 $\widetilde{f},\widetilde{f}'$ 是 f 的两个提升,并且

$$p(\Phi(\widetilde{f})) \bigcap p(\Phi(\widetilde{f}')) \neq \varnothing$$

那么存在一个升腾 $\gamma \in \mathscr{D}(\widetilde{X},p)$ 使得

$$\widetilde{f}'=\gamma \circ \widetilde{f} \circ \gamma^{-1}$$

并且

$$p(\Phi(\widetilde{f}'))=p(\Phi(\widetilde{f}))$$

对于一个给定的 $\widetilde{f}$,把

$$\{\widetilde{f}'=\gamma \circ \widetilde{f} \circ \gamma^{-1} \mid \gamma \in \mathscr{D}(\widetilde{X},p)\}$$

叫作 f 的一个提升 H 类,记作 $[\widetilde{f}]$;明显地,$[\widetilde{f}']=[\widetilde{f}]$,其中 $\widetilde{f}'=\gamma \circ \widetilde{f} \circ \gamma^{-1}$.这就把所有提升划分为两两不相交的提

升 H 类.还把 $p(\Phi(\widetilde{f}))$ 叫作 f 的一个不动点 H 类;它不依赖于 $[\widetilde{f}]$ 中所选取的代表 $\widetilde{f}$,所以可以把它记作 $p(\Phi([\widetilde{f}]))$. f 的每一个不动点属于一个不动点 H 类,并且 $\Phi(f)$ 划分为两两不相交的不动点 H 类.这就给出下述引理:

引理 1 $\Phi(f)$ 与不动点 H 类有下述关系

$$\Phi(f) = \bigcup_{[\widetilde{f}]} p(\Phi(\widetilde{f}))$$

$$[\widetilde{f}] = [\widetilde{f}'] \Rightarrow p(\Phi(\widetilde{f})) = p(\Phi(\widetilde{f}'))$$

$$[\widetilde{f}] \neq [\widetilde{f}'] \Rightarrow p(\Phi(\widetilde{f})) \bigcap p(\Phi(\widetilde{f}')) = \varnothing$$

其次,我们将进而考虑同伦.因为 $\widetilde{X}$ 是多面体 X 的有限叶的复迭空间,$\widetilde{X}$ 也是多面体,所以提升 $\widetilde{f}:\widetilde{X} \to \widetilde{X}$ 有它的 Lefschetz 数 $L(\widetilde{f})$.如果 $\widetilde{f}' = \gamma \circ \widetilde{f} \circ \gamma^{-1}, \gamma \in \mathscr{D}(\widetilde{X}, p)$,那么 $L(\widetilde{f}') = L(\widetilde{f})$.所以我们可以通过

$$L([\widetilde{f}]) = L(\widetilde{f})$$

来引进记号 $L([\widetilde{f}])$,把它叫作提升 H 类 $[\widetilde{f}]$ 的 Lefschetz 数.

设 $f_0:X \to X$ 满足式(1),并且 $\widetilde{f}_0$ 是 f_0 在 $\widetilde{X}$ 上的一个提升.如果同伦 $F:f_0 \simeq f_1:X \to X$,易知:f_1 也满足基本假设.那么,F 恰有一个提升 $\widetilde{F}$,以 $\widetilde{f}_0$ 为起点.记 $\widetilde{F}$ 的终点为 $\widetilde{f}_1$.明显地,$L(\widetilde{f}_0) = L(\widetilde{f}_1)$.则一个同伦 F 给出 f_0 的提升 H 类与 f_1 的提升 H 类之间的一个一一对应,并且对应的提升 H 类的 Lefschetz 数相等,即

$$L([\widetilde{f}_0]) = L([\widetilde{f}_1])$$

我们把这些事实简述成下面的引理:

引理 2 f 的每一提升 H 类 $[\widetilde{f}]$ 的 Lefschetz 数 $L([\widetilde{f}])$ 是同伦不变量.

定义 1　用 $N(f,H)$ 记 f 的、具有非零的 Lefschetz 数的提升 H 类 $[\tilde f]$ 的个数. 它是本节标题中所说的另一种 Nielsen 数.

定理 1　同伦于 f 的映射至少有 $N(f,H)$ 个不动点.

证明　因为引理 2，我们只需要考虑 f 本身. $L([\tilde f]) \neq 0 \Rightarrow$ 不动点 H 类 $p(\Phi(\tilde f))$ 非空. 不同的不动点 H 类是 $\Phi(f)$ 中不相交的子集.

系 1（Hirsch）　设 f 是连通的多面体 X 的自映射，并且 $p:\tilde X \to X$ 是 X 的两叶连通的复迭空间. 再设：

(i) f 具有提升 $\tilde f_1, \tilde f_2 : \tilde X \to \tilde X$；

(ii) 如果 $p(a)=p(b)$，并且 $a \neq b$，那么 $\tilde f_i(a) \neq \tilde f_i(b), i=1,2$；

(iii) $L(\tilde f_i) \neq 0, i=1,2$.

那么，每一个同伦于 f 的映射都至少有两个不动点.

证明　因为 $(\tilde X, p)$ 是两叶的，所以它是 X 的正则复迭空间，$\# \dfrac{\pi_1(X, x_0)}{H(X, x_0)} = 2$，并且 $\mathscr{D}(\tilde X, p)$ 只有两个元素，其中一个是恒同映射，另一个互换每一纤维中的两个点，记作 γ. 根据(i)，f 具有提升，所以满足式(1)，因此本节的基本假设是满足的. 根据(ii)，有

$$\gamma \circ \tilde f_i \circ \gamma^{-1}(a) = \gamma \circ \tilde f_i(b) = \tilde f_i(a)$$

即 $[\tilde f_i]$ 只有一个元素 $\tilde f_i$，即

$$[\tilde f_1] \neq [\tilde f_2]$$

再由(iii)得 $N(f,H)=2$.

最后，我们只叙述下面的重要定理而不证明.

定理 2　$N(f,H) \leqslant N(f)$.

二、例（闭流形的自同胚）

设 T_2 是有两个洞的环面（图4）. 在本节中我们先给出 T_2 的一个自同胚 f，有 $L(f)=0$，但通过应用 164 页的系 1，

证明每一个同伦于 f 的自映射都至少有两个不动点.

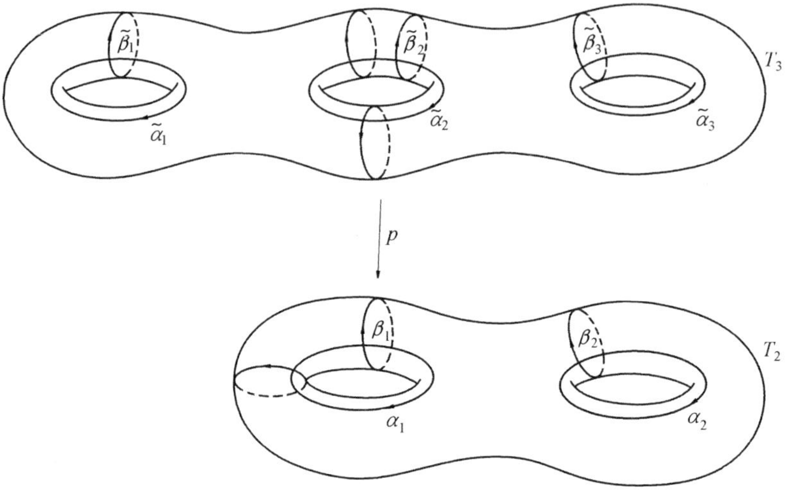

图 4

在图 4 中 $\alpha_1,\beta_1,\alpha_2,\beta_2$ 是 T_2 的一维同调群 $H_1(T_2)$ 的基. 我们所定义的自同胚 $f:T_2 \to T_2$ 是下列两个自同胚的合成. 第一个自同胚是先将 T_2 沿 β_1 剪开, 然后把剪缝左边的部分沿 β_1 上箭头所示的相反方向扭转 $360°$ 后重新粘上. 第二个自同胚是将 T_2 沿 α_1 剪开, 然后把剪缝下面沿 α_1 上箭头所示的方向扭转 $720°$ 后再重新粘上. 这两个映射对 $H_1(T_2)$ 的基 $\alpha_1,\beta_1,\alpha_2,\beta_2$ 的变换矩阵分别为

$$\begin{bmatrix} 1 & -1 & 0 & 0 \\ 0 & 1 & 0 & 0 \\ 0 & 0 & 1 & 0 \\ 0 & 0 & 0 & 1 \end{bmatrix} \text{和} \begin{bmatrix} 1 & 0 & 0 & 0 \\ 2 & 1 & 0 & 0 \\ 0 & 0 & 1 & 0 \\ 0 & 0 & 0 & 1 \end{bmatrix}$$

上述矩阵的积为

$$\begin{bmatrix} -1 & -1 & 0 & 0 \\ 2 & 1 & 0 & 0 \\ 0 & 0 & 1 & 0 \\ 0 & 0 & 0 & 1 \end{bmatrix}$$

它是在同态 f_{1*} 下 $H_1(T_2)$ 的基的变换矩阵. 又因为 f 是一个保持定向的同胚, 所以

$$f_{2*} = \mathrm{id}: H_2(T_2) \to H_2(T_2)$$

于是

$$L(f) = \sum_{i=0}^{2} (-1)^i \mathrm{tr}(f_{1*}) = 1 - 2 + 1 = 0$$

设 T_3 是带有三个洞的环面(图 4). 穿过中间那个洞的铅垂线是 T_3 的对称轴, 以下简称为轴线. 我们定义映射 $p: T_3 \to T_2$, 使 (T_3, p) 是 T_2 的一个两叶的复迭空间. 把 T_3 沿图 4 中两条无标号的闭曲线剪开为两部分, 然后将每部分中作为边界的两条闭曲线上关于轴线的对称点看作同一点, 并将这两部分同胚地映射到 T_2, 使 T_3 上看作同一曲线的那两条无标号曲线映射成 T_2 上的那条无标号曲线, 并且还保持方向一致. 容易验证, 这样定义的 p, 使 (T_3, p) 是 T_2 的一个两叶的复迭空间. 升腾群 $\mathscr{D}(T_3, p)$ 只有两个元素, 一个是恒同自映射, 另一个是绕轴线转 $180°$ 的自映射 γ.

从 $f: T_2 \to T_2$ 可以构造出 T_3 的一个自映射 $\tilde{f}$, 使 $\tilde{f}$ 是 f 的一个提升. 为此, 只要令 $\tilde{f}$ 是下列两个映射的合成即可, 第一个映射将 T_3 沿 $\tilde{\beta}_2$ 剪开, 将剪开的左边部分沿 $\tilde{\beta}_2$ 所示相反的方向扭转 $360°$ 后再粘上, 并在 $\tilde{\beta}_2$ 的对称位置上对称地做; 第二个映射是将 T_3 沿 $\tilde{\alpha}_2$ 剪开, 剪开后把剪缝下面部分沿 $\tilde{\alpha}_2$ 所示方向扭转 $360°$ 后再粘上. $\tilde{f}$ 是 f 的一个提升, 在同态 $\tilde{f}_{1*}$ 下, $H_1(T_3)$ 的基 $\tilde{\alpha}_1, \tilde{\beta}_1, \tilde{\alpha}_2, \tilde{\beta}_2, \tilde{\alpha}_3, \tilde{\beta}_3$ 的变换矩阵为

$$\begin{bmatrix} 1 & 0 & 0 & 0 & 0 & 0 \\ 0 & 1 & 0 & 0 & 0 & 0 \\ 0 & 0 & 1 & -2 & 0 & 0 \\ 0 & 0 & 0 & 1 & 0 & 0 \\ 0 & 0 & 0 & 0 & 1 & 0 \\ 0 & 0 & 0 & 0 & 0 & 1 \end{bmatrix} \cdot \begin{bmatrix} 1 & 0 & 0 & 0 & 0 & 0 \\ 0 & 1 & 0 & 0 & 0 & 0 \\ 0 & 0 & 1 & 0 & 0 & 0 \\ 0 & 0 & 1 & 1 & 0 & 0 \\ 0 & 0 & 0 & 0 & 1 & 0 \\ 0 & 0 & 0 & 0 & 0 & 1 \end{bmatrix}$$

$$= \begin{bmatrix} 1 & 0 & 0 & 0 & 0 & 0 \\ 0 & 1 & 0 & 0 & 0 & 0 \\ 0 & 0 & -1 & -2 & 0 & 0 \\ 0 & 0 & 1 & 1 & 0 & 0 \\ 0 & 0 & 0 & 0 & 1 & 0 \\ 0 & 0 & 0 & 0 & 0 & 1 \end{bmatrix}$$

于是，$L(\widetilde{f}) = 1 - 4 + 1 = -2.$

令 $\widetilde{f}' = \gamma \circ \widetilde{f}.$ $\widetilde{f}'$ 对 $H_1(T_3)$ 的基的变换矩阵为

$$\begin{bmatrix} 1 & 0 & 0 & 0 & 0 & 0 \\ 0 & 1 & 0 & 0 & 0 & 0 \\ 0 & 0 & -1 & -2 & 0 & 0 \\ 0 & 0 & 1 & 1 & 0 & 0 \\ 0 & 0 & 0 & 0 & 1 & 0 \\ 0 & 0 & 0 & 0 & 0 & 1 \end{bmatrix} \cdot \begin{bmatrix} 0 & 0 & 0 & 0 & 1 & 0 \\ 0 & 0 & 0 & 0 & 0 & 1 \\ 0 & 0 & 1 & 0 & 0 & 0 \\ 0 & 0 & 0 & 1 & 0 & 0 \\ 1 & 0 & 0 & 0 & 0 & 0 \\ 0 & 1 & 0 & 0 & 0 & 0 \end{bmatrix}$$

$$= \begin{bmatrix} 0 & 0 & 0 & 0 & 1 & 0 \\ 0 & 0 & 0 & 0 & 0 & 1 \\ 0 & 0 & -1 & -2 & 0 & 0 \\ 0 & 0 & 1 & 1 & 0 & 0 \\ 1 & 0 & 0 & 0 & 0 & 0 \\ 0 & 1 & 0 & 0 & 0 & 0 \end{bmatrix}$$

因此 $L(\widetilde{f}') = 1 - 0 + 1 = 2.$

应用 164 页的系 1 立即有：每一个同伦于 f 的映射都至少有两个不动点.

这个例中的 T_2 是二维的. 我们只要将此例稍加改造，就可以构造：一个 $n(n \geqslant 3)$ 维的闭流形 M^n 和一个同胚映射 $h: M^n \to M^n$，使得 $L(h) = 0$，而每一个与 h 同伦的映射都至少有两个不动点.

设同胚映射 $g: S^{n-2} \to S^{n-2}$，有 $L(g) \neq 0.$ 令 $M^n = T_2 \times S^{n-2}$，$h = f \times g$，则

$$h = f \times g: T_2 \times S^{n-2} \to T_2 \times S^{n-2}$$

是 n 维闭流形 $T_2 \times S^{n-2}$ 的一个自同胚，并且

$$L(h) = L(f) \cdot L(g) = 0$$

因为 $(T_3 \times S^{n-2}, p \times \mathrm{id})$ 是 $T_2 \times S^{n-2}$ 的一个两叶的复迭空间，映射

$$\tilde{f} \times g, \tilde{f}' \times g : T_3 \times S^{n-2} \to T_3 \times S^{n-2}$$

是 h 的两个提升，有

$$L(\tilde{f} \times g) = L(\tilde{f}) \cdot L(g) \neq 0$$

和

$$L(\tilde{f}' \times g) = L(\tilde{f}') \cdot L(g) \neq 0$$

所以，从 164 页的系 1，每一个同伦于 h 的映射至少有两个不动点.

三、从自映射的不动点类到方程的根类

设道路连通的拓扑空间 X 是一个拓扑群，1 是它的单位点，并且 x_* 是它的一个已知点. 那么，对于 X 的任一自映射 f，x 是方程

$$f(x) = x_* \tag{1}$$

的一个根 $\Leftrightarrow x$ 是方程

$$x_*^{-1} \cdot f(x) = 1 \text{（或 } f(x) \cdot x_*^{-1} = 1) \tag{2}$$

的一个根 $\Leftrightarrow x$ 是 X 的自映射

$$g(x) = x_*^{-1} \cdot f(x) \cdot x$$

（或 $x \cdot x_*^{-1} \cdot f(x)$，或 $f(x) \cdot x_*^{-1} \cdot x$，或 $x \cdot f(x) \cdot x_*^{-1}$ 等）

$$\tag{3}$$

的一个不动点. 这说明对于拓扑群 X，方程 (1) 的根与 X 的自映射 g 的不动点之间有密切关系.

从这种关系自然产生一个问题：不动点类这一概念是否也给出"根类"的概念？答案是肯定的. 设 x_0, x_1 是 X 的自映射 g 的两个不动点，则它们属于 g 的同一个不动点类 $\Leftrightarrow X$ 上存在一条从 x_0 到 x_1 的道路 c，使得

$$c \underset{\cdot}{\simeq} g \circ c \tag{4}$$

即存在连接 c 与 $g \circ c$ 的定端同伦

$$G : I \times I \to X$$

此外,从式(3)有

$$f(x) = x_* \cdot g(x) \cdot x^{-1} \tag{5}$$

令 $F : I \times I \to X$ 为

$$F(t, s) = x_* \cdot G(t, s) \cdot c(t)^{-1}$$

(其中 $c(t)^{-1}$ 指的是拓扑群 X 的点 $c(t)$ 的逆点,非道路 c 的逆 c^{-1}). 则有

$$F(t, 0) = x_*, F(t, 1) = x_* \cdot g \circ c(t) \cdot c(t)^{-1} = f \circ c(t)$$

和

$$F(0, s) = x_*, F(1, s) = x_*$$

这就证明了下面结果中的"⇒"部分.

设 x_0, x_1 是方程 $f(x) = x_*$ 的根. 那么, x_0, x_1 属于 $g(x) = x_*^{-1} \cdot f(x) \cdot x$ 的同一不动点类 $\Leftrightarrow X$ 上存在一条从 x_0 到 x_1 的道路 c, 使

$$f \circ c \simeq e_*$$

这里 $e_* : I \to x_*$ 是点道路.

"⇐"部分的证明相仿.

把上面"⇔"后的一句话作为"根类"的定义,我们就得到了根类的概念,并且证明了:当 X 是拓扑群时,式(3)中 g 的不动点类是方程(1)的根类,而且反之也成立;换句话说,这两者可以互译.

以上讨论只限于 X 是拓扑群. 此后我们取消这种限制,以下面的定义为出发点.

定义2 设 f 是从拓扑空间 Y 到拓扑空间 X 的任一映射,并且 x_* 是 X 的一个已知点. 把方程

$$f(y) = x_*$$

的全体根所组成的集合记作 $\Gamma(f, x_*)$, 即 $\Gamma(f, x_*) = f^{-1}(x_*)$.

两个根 y_0, y_1 称为同类的,如果 Y 中存在一条从 y_0 到 y_1 的道路 c, 使得

$$f \circ c \simeq e_* \quad 或 \langle f \circ c \rangle = e$$

这里 $e_*: I \to x_*$ 是点道路,e 是 $\pi_1(X, x_*)$ 的单位元.

容易看出根的同类关系是一个等价关系,从而把 $\Gamma(f, x_*)$ 划分为两两不相交的子集.这些子集的每一个叫作一个根类.把以根类为元素的集合记作 $\Gamma'(f, x_*)$.(注意每一个根类都是非空集.)

引理 3 如果 X 是 T_1 空间,那么 $\Gamma(f, x_*)$ 是闭的.

证明 从假设,单点集 $\{x_*\}$ 是 X 中的闭集,因此 $\Gamma(f, x_*) = f^{-1}(x_*)$ 是 Y 中的闭集.

引理 4 如果 Y 是局部道路连通的,X 是局部单连通的,那么任一根类 $\mathcal{R} \in \Gamma'(f, x_*)$ 是 $\Gamma(f, x_*)$ 中的一个开集.

证明 设 $y \in \mathcal{R}$.在 X 中取 x_* 的一个邻域 V,使它满足:如果 c 是 V 中 x_* 处的一个圈,就有 $\langle c \rangle = \langle e_* \rangle$.$f^{-1}(V)$ 是 y 的一个邻域,于是 $f^{-1}(V)$ 包含 y 的一个道路连通的邻域 U.我们证明,如果 $y' \in U \bigcap \Gamma(f, x_*)$,那么 $y' \in \mathcal{R}$.事实上,因为 U 是道路连通的,所以在 U 中存在一条从 y 到 y' 的道路 d.这时 $f \circ d$ 是在 V 中,并且是 x_* 处的一个圈,因此有 $\langle f \circ d \rangle = \langle e_* \rangle$.这表明,$U \bigcap \Gamma(f, x_*) \subseteq \mathcal{R}$.所以 $\mathcal{R}$ 是 $\Gamma(f, x_*)$ 中的开集.

定理 3 如果 Y 是紧致的,局部道路是连通的,X 是局部单连通的 T_1 空间,那么根类的集合 $\Gamma'(f, x_*)$ 是有限的,并且每一个根类是紧致的.

证明 从引理 3,$\Gamma(f, x_*)$ 是闭的,又 Y 是紧致的,所以 $\Gamma(f, x_*)$ 是紧致的.根据引理 4 有本定理的结论.

例 1 如果 A 是 Y 的一个道路连通的非空子集,并且它的点都是方程 $f(y) = x_*$ 的根,那么 A 就属于同一个根类.如果 $f: Y \to x_0$ 是常映射,按照 $x_0 = x_*$ 或 $x_0 \neq x_*$,分别有 $\Gamma'(f, x_*)$ 的元素就是 Y 的诸道路连通分支,或 $\Gamma(f, x_*) = \varnothing$.

例 2 如果 $f: \mathbb{R}^1 \to \mathbb{R}^1, x \mapsto x^2$,那么 $\Gamma(f, 1) = \{-1,$

1},并且这两个根属于同一个根类,而 $\Gamma(f,-1)=\varnothing$. $\mathbb{R}^1$ 在加法运算下是一个拓扑群. 当 $x_*=1$ 时式(3)中的 $g(x)=x^2+x-1$.

例 3　考虑圆周 S^1 的整幂映射 $f_n:z\mapsto z^n(n\neq 0)$ 与方程 $f_n(z)=1$. 那么,此方程有 $|n|$ 个根,即

$$\Gamma(f_n,1)=\{\mathrm{e}^{\frac{2\pi ri}{|n|}},r=1,2,\cdots,|n|\}$$

并且每一个根单独地组成一个根类. 圆周在复数的乘法运算下是一个拓扑群,式(3)中的 $g(z)=z^{n+1}$.

后两例中的 $Y=X$ 都是拓扑群.

四、根类在映射的同伦下的对应

定义 3　设 H 是从 f_0 到 f_1 的一个同伦,$H:f_0\simeq f_1:$ $Y\to X$,x_* 是 X 的一个已知点,y_i 是方程

$$f_i(y)=x_*$$

的根,这里的 $i=0,1$. 如果 Y 中存在一条从 y_0 到 y_1 的道路 c,使得

$$\langle\Delta(H,c)\rangle=\langle c_*\rangle$$

我们就说同伦 H 给出从根 y_0 到根 y_1 的一个对应,或者说根 y_0 在同伦 H 下对应于根 y_1,记作

$$y_0Hy_1$$

容易看出 $y_0Hy_1\Leftrightarrow y_1H^{-1}y_0$.

附记 1　如果定义 3 的 H 是常同伦 $H(\cdot,t)=f$,那么定义 3 就是定义 2. 换句话说,方程 $f(y)=x_*$ 的一个根 y_0 根据定义 3 在常同伦 f 下对应于该方程的另一个根 y_1(反之,y_1 也在常同伦 f 下对应于 y_0),当且仅当 y_0 与 y_1 属于同类.

其次,如果定义 3 中的 H 是从 f 到 f 的一个闭同伦,并且 y_0 与 y_1 是同一个方程 $f(y)=x_*$ 的根,从 y_0Hy_1 不能推断 y_0 与 y_1 同类,也不能推断 y_1Hy_0.

附记 2　对于映射 $f_0,f_1:Y\to X$,即使存在同伦 $H:$ $f_0\simeq f_1$,可能 $f_0(y)=x_*$ 有根,而 $f_1(y)=x_*$ 无根;也可能 $f_i(y)=x_*$ 有根 y_i,$i=0,1$,但在同伦 H 下 y_0 不对应于 y_1. 这两种情形见下面的两例.

例1 设 $H(x,t) = x^2 + 2t$ 定义了从 $f_0(x) = x^2$ 到 $f_1(x) = x^2 + 2$ 的一个同伦 $H: f_0 \simeq f_1: \mathbb{R}^1 \to \mathbb{R}^1$. $\Gamma(f_0, 1) = \{-1, 1\}$,而 $\Gamma(f_1, 1) = \varnothing$. 所以 $f_0(x) = x^2 = 1$ 的根 -1 或 1 在同伦 H 下不对应于 $f_1(x) = x^2 + 2 = 1$ 的根,因为后一方程无实根.

例2 设 $H(z,t) = z^2 \mathrm{e}^{-\pi t i}$ 定义了从 $f_0(z) = z^2$ 到 $f_1(z) = z^2 c^{-\pi i}$ 的一个同伦 $H: f_0 \simeq f_1: S^1 \to S^1$,这里 S^1 是圆周. 显然有

$$\Gamma(f_0, t) = \{1, -1\}, \Gamma(f_1, 1) = \{\mathrm{e}^{\frac{\pi i}{2}}, \mathrm{e}^{\frac{3\pi i}{2}}\}$$

取 $c(t) = \mathrm{e}^{\frac{\pi t i}{2}}$ 为 S^1 上从 $\Gamma(f_0, 1)$ 中的根 1 到 $\Gamma(f_1, 1)$ 中的根 $\mathrm{e}^{\frac{\pi i}{2}}$ 的道路,那么 $\Delta(H, c)(t) \equiv 1$,所以在 H 下 $\Gamma(f_0, 1)$ 中的根 1 对应于 $\Gamma(f_1, 1)$ 中的根 $\mathrm{e}^{\frac{\pi i}{2}}$. 设 c 为 S^1 上从 1 到 $\mathrm{e}^{\frac{3\pi i}{2}}$ 的任一道路. 那么,一定存在某一整数 k,使

$$c \simeq c_k$$

其中 $c_k(t) = \mathrm{e}^{\left(2k + \frac{3}{2}\right)\pi t i}, \forall t \in I$. 而 $\Delta(H, c_k)(t) = \mathrm{e}^{(4k+2)\pi t i}$,于是

$$\langle \Delta(H, c) \rangle = \langle \Delta(H, c_k) \rangle \neq \langle c_* \rangle$$

即 $\Gamma(f_0, 1)$ 中的根 1 在同伦 H 下不对应于 $\Gamma(f_1, 1)$ 中的根 $\mathrm{e}^{\frac{3\pi i}{2}}$.

还容易看出 $\Gamma'(f_0, 1)$ 与 $\Gamma'(f_1, 1)$ 各有两个元素,并且在同伦 H 下一一对应.

定理4 设存在同伦 $H: f_0 \simeq f_1: Y \to X, y_i \in \Gamma(f_i, x_*), i = 0, 1$,并且 $y_0 H y_1$. 记 $y_i \in$ 根类 $\mathscr{R}_i \in \Gamma'(f_i, x_*), i = 0, 1$. 那么:

(i) $y'_0 \in \mathscr{R}_0 \Leftrightarrow y'_0 H y_1$;

(ii) $y'_1 \in \mathscr{R}_1 \Leftrightarrow y_0 H y'_1$.

换句话说,$y_0 H y_1$ 诱导出在同伦 H 下根类 $\mathscr{R}_0$ 对应于根类 $\mathscr{R}_1$,记作

$$\mathscr{R}_0 H \mathscr{R}_1$$

容易看出 $\mathscr{R}_0 H \mathscr{R}_1 \Leftrightarrow \mathscr{R}_1 H^{-1} \mathscr{R}_0$.

证明　(i) 中的"⇒". 从假设 $y_0, y'_0 \in \mathcal{R}_0 \Rightarrow$ 存在从 y'_0 到 y_0 的道路 c_0, 使得 $\langle \Delta(f_0, c_0) \rangle = \langle e_* \rangle$, 其中 f_0 看作常同伦. $y_0 H y_1 \Rightarrow$ 存在从 y_0 到 y_1 的道路 c, 使得 $\langle \Delta(H, c) \rangle = \langle e_* \rangle$. 但

$$H \simeq f_0 H$$

从而

$$\begin{aligned}
\langle \Delta(H, c_0 c) \rangle &= \langle \Delta(f_0 H, c_0 c) \rangle \\
&= \langle \Delta(f_0, c_0) \rangle \langle \Delta(H, c) \rangle \\
&= \langle e_* \rangle
\end{aligned}$$

(i) 中的"⇐". 从假设 $y_0 H y_1$ 和 $y'_0 H y_1$, 即 $y_1 H^{-1} y'_0 \Rightarrow$ 存在从 y_0 到 y_1 的道路 c 与从 y_1 到 y'_0 的道路 c', 使得 $\langle \Delta(H, c) \rangle = \langle \Delta(H^{-1}, c') \rangle = \langle e_* \rangle$. 又

$$\begin{aligned}
\langle \Delta(f_0, c c') \rangle &= \langle \Delta(H H^{-1}, c c') \rangle \\
&= \langle \Delta(H, c) \rangle \langle \Delta(H^{-1}, c') \rangle \\
&= \langle e_* \rangle
\end{aligned}$$

(ii) 由 $y_0 H y'_1 \Rightarrow y'_1 H^{-1} y_0$, 再利用 (i) 立即可得.

定义 4　设 $f: Y \to X$, 并有同伦 $H: f \simeq H(\cdot, 1): Y \to X$. 如果方程

$$f(y) = x_* \tag{1}$$

的一个根类 $\mathcal{R}(\in \Gamma'(f, x_*))$ 在任意这样的一个同伦 H 下都对应于一个根类 $\in \Gamma'(H(\cdot, 1), x_*)$, 那么就说根类 $\mathcal{R}$ 是本质的. 以本质的根类为元素的集合记作 $\Gamma^*(f, x_*)$; 集合 $\Gamma^*(f, x_*)$ 中的元素个数 $\sharp \Gamma^*(f, x_*)$ 叫作方程 (1) 的 Nielsen 数, 记作 $N(f, x_*)$.

定理 5　如果 $\mathcal{R}$ 是 $\Gamma^*(f, x_*)$ 中本质的根类, 那么, 在以 f 为起点的任一同伦 H 下, $\mathcal{R}$ 恰对应于一个本质的根类 $\mathcal{R}_1 \in \Gamma^*(H(\cdot, 1), x_*)$.

从而, 在同伦 H 和 H^{-1} 下, 根的对应 $y_0 H y_1$ 和 $y_1 H^{-1} y_0$ (定义 3) 诱导出 $\Gamma^*(f, x_*)$ 与 $\Gamma^*(H(\cdot, 1), x_*)$ 之间的一对互逆的一一对应. 于是 Nielsen 数相等

$$N(f,x_*) = N(H(\cdot,1),x_*)$$

即 Nielsen 数 $N(f,x_*)$ 有同伦不变性.

证明 从假设, $\mathscr{R}$ 是本质的根类. 根据定义 4, 有一个根类 $\mathscr{R}_1 \in \Gamma'(H(\cdot,1),x_*)$, 使得 $\mathscr{R}H\mathscr{R}_1$. 再根据定理 4, 这个 $\mathscr{R}_1$ 是唯一的, 所以现在必须证明 $\mathscr{R}_1$ 也是本质的根类. 因此, 如果 H' 是以 $H(\cdot,1)$ 为起点的任一同伦, 那么必须证明存在 $\mathscr{R}_2(\in \Gamma'(H'(\cdot,1),x_*))$ 使得 $\mathscr{R}_1 H' \mathscr{R}_2$. 现在, 既然 $\mathscr{R}$ 是本质的, 那么在同伦 HH' 下必然存在 $\mathscr{R}_2 \in \Gamma'(H'(\cdot,1),x_*)$, 使得 $\mathscr{R}(HH')\mathscr{R}_2$; 从 $\mathscr{R}_1 H^{-1}\mathscr{R}$ 和 $\mathscr{R}(HH')\mathscr{R}_2$, 给出 $\mathscr{R}_1 H^{-1}(HH')\mathscr{R}_2$, 因而 $\mathscr{R}_1 H' \mathscr{R}_2$. 这就是说 $\mathscr{R}_1$ 也是本质的.

五、X 的基本群 $\pi_1(X,x_*)$ 的另一个子群 $S(X,x_*)$

目前暂时只考虑一个拓扑空间 X, 而不涉及前两部分中的空间 Y 与映射 $f:Y \to X$.

定义 5 设 X 是拓扑空间, x_* 是其中一个已知点. 设 $H:X \times I \to X$ 是映射, 并且 $H(\cdot,0)$ 与 $H(\cdot,1)$ 都是以 x_* 为不动点的同胚. 从而 $\Delta(H,x_*)$[①] 是 X 中点 x_* 处的一个圈. 对于所有这种 H, 把定端同伦类 $\langle \Delta(H,x_*)\rangle$ 的全体记作 $S(X,x_*)$.

定理 6 $S(X,x_*)$ 是 $\pi_1(X,x_*)$ 的一个子群. 明显地, J 群 $J(\mathrm{id},x_*) \subseteq S(X,x_*)$.

为了证明本定理, 先证明下面的引理.

引理 5 如果 $\alpha \in S(X,x_*)$, 那么, 存在一个同伦 $H:X \times I \to X$, 使得 $H(\cdot,0)$ 是同胚, 以 x_* 为不动点, $H(\cdot,1)$ 是 X 的恒同映射 id_X, 并且 $\langle \Delta(H,x_*)\rangle = \alpha$.

证明 由于 $\alpha \in S(X,x_*)$, 因此存在一个同伦 $H':X \times I \to X$, 有 $H'(\cdot,0)$ 和 $H'(\cdot,1)$ 都是以 x_* 为不动点的同胚, 并且 $\langle \Delta(H',x_*)\rangle = \alpha$. 将同胚映射 $H'(\cdot,1)$ 的逆映射记作 g, 有 $H'(\cdot,1) \circ g = \mathrm{id}_X$. 令 $H:X \times I \to X$ 为

---

① $\Delta(H,x_*)$ 中 x_* 表示点 x_* 处的点道路.

$H(x,t) = H'(g(x),t)$. 从 $g(x_*) = x_*$, 有 $\Delta(H,x_*) = \Delta(H',x_*)$. 则同伦 H 就是本引理所要求的同伦.

定理 6 的证明　设 $\alpha_1,\alpha_2 \in S(X,x_*)$, 我们证明 $\alpha_1\alpha_2^{-1} \in S(X,x_*)$. 根据引理 5, 存在同伦 $H_i : X \times I \to X$, 有 $H_i(\cdot,0)$ 是同胚, 以 x_* 为不动点, $H_i(\cdot,1) = \mathrm{id}_X$, 并且 $\langle\Delta(H_i,x_*)\rangle = \alpha_i, i = 0,1$. 令 $H = H_1 H_2^{-1}$. 则

$$\begin{aligned}
\alpha_1\alpha_2^{-1} &= \langle\Delta(H_1,x_*)\rangle\langle\Delta(H_2,x_*)\rangle^{-1} \\
&= \langle\Delta(H_1 H_2^{-1},x_*)\rangle \\
&= \langle\Delta(H,x_*)\rangle \in S(X,x_*)
\end{aligned}$$

定理 7　如果 X 是一个流形①, 那么

$$S(X,x_*) = \pi_1(X,x_*)$$

证明　对于任一元素 $\langle c\rangle \in \pi_1(X,x_*)$, 我们证明: 存在同伦 $H : X \times I \to X$, 有 $H(\cdot,0)$ 和 $H(\cdot,1)$ 是以 x_* 为不动点的同胚, 并且 $\langle\Delta(H,x_*)\rangle = \langle c\rangle$.

由于 X 是流形, 因此存在开集 $U_i \subset X$, 同胚 $h_i : U_i \to \mathbb{R}^n$ 和道路 $c_i : I \to X, i = 1,\cdots,m$, 使 $c = c_1 c_2 \cdots c_m$, 并且 $c_i(I) \subset U_i, h_i(c_i(I))$ 在单位球内部, $h_i(c_i(1)) = 0, i = 1,\cdots,m$. 对于每个 i 定义同伦 $H_i : X \times I \to X$ 为 (图 5)

$$H_i(x,t) = \begin{cases} h_i^{-1}(h_i(x) + (1 - |h_i(x)|)h_i(c_i(t))), & \text{当 } x \in U_i \\ & \text{并且 } |h_i(x)| \leqslant 1 \\ x, & \text{对其余的 } x \end{cases}$$

则 $\Delta(H_i,c_i(1))(t) = H_i(c_i(1),t) = h_i^{-1}(h_i(c_i(t))) = c_i(t)$, 并且 $H_i(\cdot,0)$ 是同胚, $H_i(\cdot,1) = \mathrm{id}_X, i = 1,\cdots,m$.

现在令

$$H'_m = H_m, \quad H'_i(x,t) = H_i(H'_{i+1}(x,0),t), i < m$$

对于每一个 $i < m$, 有

$$H'_i(x,1) = H_i(H'_{i+1}(x,0),1) = H'_{i+1}(x,0)$$

---

①　如果一个 Hausdorff 空间 X 中的每一点有一个与 $\mathbb{R}^n$ 同胚的邻域, 那么称 X 为 n 维的流形.

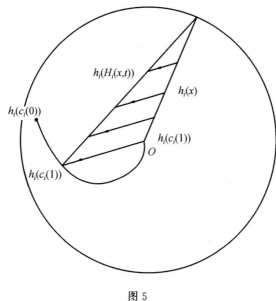

图 5

而

$$\Delta(H'_m, x_*) = \Delta(H_m, x_*) = c_m$$

于是

$$\begin{aligned}
\Delta(H'_{m-1}, x_*)(t) &= H_{m-1}(H'_m(x_*,0),t) \\
&= H_{m-1}(c_m(0),t) \\
&= H_{m-1}(c_{m-1}(1),t) \\
&= c_{m-1}(t)
\end{aligned}$$

即 $\Delta(H'_{m-1}, x_*) = c_{m-1}$, 依此类推, 对 $i < m$, 有 $\Delta(H'_i, x_*) = c_i$.

定义同伦 $H: X \times I \to X$ 为 $H = H'_1 H'_2 \cdots H'_m$, 则

$$\begin{aligned}
\langle \Delta(H, x_*) \rangle &= \langle \Delta(H'_1 H'_2 \cdots H'_m, x_*) \rangle \\
&= \langle \Delta(H'_1, x_*) \rangle \cdots \langle \Delta(H'_m, x_*) \rangle \\
&= \langle c_1 \rangle \cdots \langle c_m \rangle = \langle c \rangle
\end{aligned}$$

六、方程的 Reidemeister 数

我们的方程

$$f(y) = x_*$$

是由道路连通的拓扑空间之间的映射 $f:Y \to X$ 与一个已知点 $x_* \in X$ 确定的. 设 $y_* \in \Gamma(f, x_*)$, 那么 f 诱导出同态

$$f_\pi : \pi_1(Y, y_*) \to \pi_1(X, x_*)$$

从而有 $\mathrm{Im}\, f_\pi$. 把它在 $\pi_1(X, x_*)$ 中右傍系的集合记作 Cosets $\mathrm{Im}\, f_\pi$, 并且把 Cosets $\mathrm{Im}\, f_\pi$ 中元素的个数叫作方程的 Reidemeister 数, 记作 $R(f, x_*)$, 即

$$R(f, x_*) = \# \mathrm{Cosets}\, \mathrm{Im}\, f_\pi$$

附记 1 设 $y_* \in \mathscr{R}_* \in \Gamma'(f, x_*)$. $\mathrm{Im}\, f_\pi$ 与 Cosets $\mathrm{Im}\, f_\pi$ 不依赖于 $\mathscr{R}_*$ 中 y_* 的选取, 而 $R(f, x_*)$ 还不依赖于 $\mathscr{R}_*$.

下面的定理建立了根类与 $R(f, x_*)$ 之间的联系.

定理 8 取定一个根类 $\mathscr{R}_* \in \Gamma'(f, x_*)$. 对任意 $\mathscr{R}_1 \in \Gamma'(f, x_*)$, 取 $y_* \in \mathscr{R}_*$, $y_1 \in \mathscr{R}_1$ 以及 Y 中从 y_* 到 y_1 的一条道路 c, 定义对应

$\varphi : \Gamma'(f, x_*) \to \mathrm{Cosets}\, \mathrm{Im}\, f_\pi, \mathscr{R}_1 \mapsto \varphi(\mathscr{R}_1) = (\mathrm{Im}\, f_\pi) \langle f \circ c \rangle$

这个对应 φ 是单值的 (即不依赖于道路 c 的选取, 不依赖于 $\mathscr{R}_*$ 中 y_* 的选取, 不依赖于 $\mathscr{R}_1$ 中 y_1 的选取), 而且还是单一的. 于是 $\# \Gamma'(f, x_*) \leqslant R(f, x_*)$.

证明 首先, 证明对应 φ 是单值的.

设 c' 是从 y_* 到 y_1 的另一道路. 那么, $c' = bc$, 这里 b 是 y_* 处的一个圈. 然后 $\langle f \circ c' \rangle = \langle f \circ b \rangle \langle f \circ c \rangle \in \mathrm{Im}\, f_\pi \langle f \circ c \rangle$, 即

$$\mathrm{Im}\, f_\pi \langle f \circ c' \rangle = \mathrm{Im}\, f_\pi \langle f \circ c \rangle$$

再设从 $\mathscr{R}_1$ 中选取的是另一根 y'_1. 既然 y_1 与 y'_1 同属于 $\mathscr{R}_1$, Y 中存在从 y_1 到 y'_1 的道路 b', 使得 $\langle f \circ b' \rangle = \langle e_* \rangle$. 取从 y_* 到 y'_1 的道路 $c' = cb'$, 有 $\langle f \circ c' \rangle = \langle f \circ c \rangle$. 因而

$$\mathrm{Im}\, f_\pi \langle f \circ c' \rangle = \mathrm{Im}\, f_\pi \langle f \circ c \rangle$$

类似地可以证明 φ 与 $\mathscr{R}_*$ 中 y_* 的选取无关.

其次, 为了证明对应 φ 是单一的, 设 $\mathscr{R}_1, \mathscr{R}'_1 \in \Gamma'(f, x_*)$, 并且 $\varphi(\mathscr{R}_1) = \varphi(\mathscr{R}'_1)$. 设 $y_1 \in \mathscr{R}_1$, $y'_1 \in \mathscr{R}'_1$, 并且 c_1, c'_1 分别是从 y_* 到 y_1, y'_1 的道路. 所以

$$\varphi(\mathscr{R}_1) = \varphi(\mathscr{R}'_1) \Rightarrow \mathrm{Im}\, f_\pi \langle f \circ c_1 \rangle = \mathrm{Im}\, f_\pi \langle f \circ c'_1 \rangle$$

这推出 $\langle f \circ c'_1 \rangle \in \mathrm{Im}\, f_\pi \langle f \circ c_1 \rangle \Rightarrow Y$ 中存在 y_* 处的一个圈 b,使得

$$\langle f \circ c'_1 \rangle = \langle f \circ b \rangle \langle f \circ c_1 \rangle = \langle f \circ (bc_1) \rangle$$

但这又推出 $\langle f \circ (c'^{-1}_1 bc_1) \rangle = \langle e_* \rangle$. 这就是说 y_1, y'_1 属于同一个根类,即 $\mathscr{R}_1 = \mathscr{R}'_1$.

引理 6 设 $\mathscr{R}_*$ 是本质的根类,即 $\mathscr{R}_* \in \Gamma^*(f, x_*)$. 如果 $\alpha \in S(X, x_*)$,那么,存在本质的根类 $\mathscr{R}_1 \in \Gamma^*(f, x_*)$,使得

$$\varphi(\mathscr{R}_1) = (\mathrm{Im}\, f_\pi)\alpha$$

证明 (1) 因为 $\alpha \in S(X, x_*)$,根据引理 5,存在 X 的自映射的一个同伦 H,使得 $H(\cdot, 0)$ 是以 x_* 为不动点的同胚,$H(\cdot, 1) = \mathrm{id}_X$,并且 $\langle \Delta(H, x_*) \rangle = \alpha$.

(2) 用 $F(y, t) = H(f(y), t), y \in Y, t \in I$ 来定义同伦 $F: H(\cdot, 0) \circ f \simeq f$. 取根 $y_* \in \mathscr{R}_* \in \Gamma^*(f, x_*)$,根据定理 5,恰存在 $H(\cdot, 0) \circ f(y) = x_*$ 的一个本质根类 $\mathscr{R}_1 \in \Gamma^*(H(\cdot, 0) \circ f, x_*)$,使得 $\mathscr{R}_1 F \mathscr{R}_*$. 取根 $y_1 \in \mathscr{R}_1 \in \Gamma^*(H(\cdot, 0) \circ f, x_*)$,那么 $y_1 F y_*$.

(3) 现在 $H(\cdot, 0)$ 是以 x_* 为不动点的同胚这一事实 $\Rightarrow H(\cdot, 0) \circ f(y) = x_*$ 的根和 $f(y) = x_*$ 的根完全相同,并且 $f(y) = x_*$ 的含有 y_1 的根类,也就是上文的 $\mathscr{R}_1$. 我们还需要证明的是:它就是本引理求证其存在的本质根类,即下列两步:(4) 根类 $\mathscr{R}_1$ 对于 $f(y) = x_*$ 来说是本质的,即 $\mathscr{R}_1 \in \Gamma^*(f, x_*)$;(5) $\varphi(\mathscr{R}_1) = \mathrm{Im}\, f_\pi \cdot \alpha$.

(4) 为了证明 $\mathscr{R}_1 \in \Gamma^*(f, x_*)$,对于任一同伦 $G: f \simeq G(\cdot, 1): Y \to X$,根据定义 4,我们必须求出一个根 $y_2 \in \Gamma(G(\cdot, 1), x_*)$,使得 $y_1 G y_2$. 为此,用 $G'(\cdot, t) = H(\cdot, 0) \circ G(\cdot, t)$ 来定义同伦 G',它是以 $H(\cdot, 0) \circ f$ 为起点的. 既然从上文的 (2),$y_1 \in \mathscr{R}_1 \in \Gamma^*(H(\cdot, 0) \circ f, x_*)$,根据定义 4,必然有 $G'(y, 1) = x_*$ 的一个根 y_2,使得 $y_1 G' y_2$,即 Y 中存在从 y_1 到 y_2 的一条道路 c,使得 $\langle \Delta(G', c) \rangle = \langle e_* \rangle$.

因为

$$\langle \Delta(G,c) \rangle = \langle \Delta(H(\cdot,0)^{-1} \circ G',c) \rangle$$
$$= H(\cdot,0)_\pi^{-1} \langle \Delta(G',c) \rangle$$
$$= H(\cdot,0)_\pi^{-1} \langle e_* \rangle = \langle e_* \rangle$$

于是得到所求证的事实 $y_1 G y_2$.

（5）从上面的（2），$y_1 F y_*$，因而 Y 中存在从 y_1 到 y_* 的一条道路 D，使得 $\langle \Delta(F,D) \rangle = \langle e_* \rangle$.（1）中已有 $\alpha = \langle \Delta(H, x_*) \rangle$. 但 $F(y,t) = H(f(y),t)$ 以及 $f(y_*) = x_* \Rightarrow \langle \Delta(H, x_*) \rangle = \langle \Delta(F,y_*) \rangle$. 又

$$\langle \Delta(F,y_*) \rangle = \langle \Delta(Ff,DD^{-1}) \rangle$$
$$= \langle \Delta(F,D) \rangle \langle \Delta(f,D^{-1}) \rangle$$
$$= \langle e_* \rangle \langle f \circ D^{-1} \rangle$$
$$= \langle f \circ D^{-1} \rangle$$

所以

$$\alpha = \langle f \circ D^{-1} \rangle$$

这里 D^{-1} 是 Y 中从 y_* 到 y_1 的道路，根据 φ 的定义，即

$$\varphi(\mathscr{R}_1) = \operatorname{Im} f_\pi \cdot \alpha$$

定义 6   设 $f\colon (Y,y_*) \to (X,x_*)$. 从而有 Cosets $\operatorname{Im} f_\pi$（见前文）. 如果 $\alpha \in S(X,x_*)$，我们就说 Cosets $\operatorname{Im} f_\pi$ 中的 $\operatorname{Im} f_\pi \cdot \alpha$ 有一个代表在 $S(X,x_*)$ 中. 于是 $S(X,x_*)$ 中的每一元素都是某一右傍系的代表，并且根据它们是否是同一右傍系的代表把 $S(X,x_*)$ 的全体元素进行分类，把 $S(X,x_*)$ 中这样分成的所有的类的个数记作 $s(f,y_*)$.

附记 2   数 $s(f,y_*)$ 不但与 f 有关，而且与 y_* 有关，因为 $\operatorname{Im} f_\pi$ 与 Cosets $\operatorname{Im} f_\pi$ 都如此，参看附记 1.

根据定义 6 与定义 4，从定理 8 与引理 6 有：

定理 9   如果 y_* 是属于方程 $f(y) = x_*$ 的一个本质根类，那么

$$s(f,y_*) \leqslant N(f,x_*)$$

系 2   如果 $S(X,x_*) = \pi_1(X,x_*)$，且 $f(y) = x_*$ 至少有一个本质根类，那么它的每一个根类都是本质的，并且

$$N(f,x_*)=R(f,x_*).$$

证明 从定理 9 和定理 8 有

$$s(f,y_*)\leqslant N(f,x_*)\leqslant \#\Gamma'(f,x_*)\leqslant R(f,x_*)$$

从 $S(X,x_*)=\pi_1(X,x_*)$，根据 $s(f,y_*)$ 与 $R(f,x_*)$ 的定义，有

$$s(f,y_*)=R(f,x_*)$$

于是

$$N(f,x_*)=\#\Gamma'(f,x_*)=R(f,x_*)$$

根据定理 7 和系 2 立即有：

系 3 如果 Y 是一个流形，且 $f(y)=x_*$ 至少有一个本质根类，那么它的每一个根类都是本质的，并且 $N(f,x_*)=R(f,x_*)$.

七、根类的指数，$S(X,x_*)$ 最大时的 Nielsen 数的计算

定理 8 与定理 9 只给出了方程的 Nielsen 数 $N(f,x_*)$ 的一个上界与一个下界. 本节将进一步运用 $\pi_1(X,x_*)$ 的子群 $S(X,x_*)$（如同 §2 中运用 J 群），来得到与 §2 中相仿的、$N(f,x_*)$ 的估计（定理 10 与系 4）.

推导本节的结果时，关键的工具仍是指数概念. 但现在讨论的对象是道路连通的拓扑空间，必须用奇异同调论，而其他部分都只限于用单纯复形及其同调论. 为了不使篇幅过长，我们将只阐述结果而略去证明.

设拓扑空间 Y 是紧致的、正规的、局部道路连通的和道路连通的，X 是局部单连通和道路连通的 Hausdorff 空间，那么方程 $f(y)=x_*$ 的任一根类 $\mathscr{R}$ 在 Y 中有一个开邻域 U，其闭包 $\overline{U}$ 不包含其他的根（$\in\Gamma(f,x_*)$）. 考虑映射

$$Y \xrightarrow{\ i\ } (Y,Y-\mathscr{R}) \xleftarrow{\ e\ } (\overline{U},\overline{U}-\mathscr{R}) \xrightarrow{\ f'\ } (X,X-x_*)$$

这里 i,e 都是包含映射，e 又是一个切除，并且 $f'=f\,|\,\overline{U}$. 它们诱导出同态

$$f'_*\circ e_*^{-1}\circ i_*:H_*(Y)\to H_*(X,X-x_*)$$

这里 H_* 表示空间的各维奇异同调群之和. 再考虑映射

$$Y \xrightarrow{\ f\ } X \xrightarrow{\ j\ } (X, X - x_*)$$

这里 j 也是包含映射. 它们诱导出同态

$$j_* \circ f_* : H_*(Y) \to H_*(X, X - x_*)$$

定义全体根 $\Gamma(f, x_*)$ 的指数为

$$v(f, x_*) = j_* \circ f_*$$

根类 $\mathscr{R}$ 的指数为

$$v(f, x_* ; \mathscr{R}) = f'_* \circ e_*^{-1} \circ i_*$$

注意, 我们这里的指数是同态.

可以证明, 这里的指数定义也具有指数的局部理论中的通常性质, 特别地, 有下列可加性与同伦不变性

$$v(f, x_*) = \sum_{\mathscr{R} \in \Gamma'(f, x_*)} v(f, x_* ; \mathscr{R})$$

在同伦 $H : f \simeq f'$ 下, 如果 f 的根类 $\mathscr{R}$ 在 H 下对应于 f' 的根类 $\mathscr{R}'$, 那么

$$v(f, x_* ; \mathscr{R}) = v(f', x_* ; \mathscr{R}')$$

如果 f 的根类 $\mathscr{R}$ 不对应于 f 的任一根类, 那么

$$v(f, x_* ; \mathscr{R}) = 0$$

现在我们能叙述下面的定理与系, 前者是关于一般的 Y 与 X 的, 后者是关于流形的.

定理 10　如果 $j_* \circ f_* \neq 0$, 那么 $N(f, x_*) > 0$. 如果还有 $S(X, x_*) = \pi_1(X, x_*)$ 并且 j_* 是满同态, 那么每一根类 $(\in \Gamma'(f, x_*))$ 都是本质的, $N(f, x_*) = R(f, x_*)$, 并且所有根类的指数都相等, 即对于任一根类 $\mathscr{R}$ 有

$$v(f, x_*) = R(f, x_*) \cdot v(f, x_* ; \mathscr{R})$$

定义 7　设 X 与 Y 都是 n 维的可定向的紧致流形, 分别取它们的 n 维整系数同调群的母元 μ 与 ν, 使得

$$f_{*n}(\nu) = (\deg f) \mu$$

这时候, 有

$$j_{*n} : H_n(X) \to H_n(X, X - x_*)$$

是同构, 从而, 对于根类 $\mathscr{R} \in \Gamma'(f, x_*)$, 如果

$$v(f,x_*;\mathscr{R})(\nu)=m(\mathscr{R})\mu$$

那么就把 $m(\mathscr{R})$ 叫作 $\mathscr{R}$ 的重数.

系 4   如果 X 与 Y 是同维的可定向的紧致流形, 并且 $\deg f \neq 0$, 那么, 每一个根类都是本质的, 所以

$$N(f,x_*)=R(f,x_*)$$

所有根类有相等的重数, 并且所有根类的重数之和等于 $\deg f$.

近年在此领域我国学者亦有建树. 如对于三维幂零流形上的所有映射, 西江大学数学系的李宗范和首都师范大学数学系的赵学志两位教授 2008 年给出了完整计算 Nielsen 型数 $NP_n(f)$ 和 $N\Phi_n(f)$ 的显式公式. 最一般的情形已被 Heath 和 Keppelmann 讨论过, 他们研究了剩余的部分, 而在三维幂零流形映射的同伦最小周期集的研究中, 给出了三维幂零流形上所有映射的最小周期集的完整描述, 并包含了对 Jezierski 和 Marzantowicz 结果的改正. 其具体内容如下[①]:

§1 引 言

给定一个映射 $f:X \to X$, 在不动点理论中有两个著名的不变量, 它们是 Lefschetz 数 $L(f)$ 和 Nielsen 数 $N(f)$. 当 $L(f) \neq 0$ 时, 映射 f 必定有不动点, 而 $N(f)$ 是 f 所在的同伦类中映射的不动点个数的一个下界, 因此, 它比 $L(f)$ 提供了更多的信息. 对于周期点, 姜伯驹引进了两个 Nielsen 型数 $NP_n(f)$ 和 $N\Phi_n(f)$, 它们分别是最小周期恰为 n 的周期点数和周期为 n 的周期点数的下界.

很明显, 在描述自映射的周期点中, 这些 Nielsen 型数比 Lefschetz 数更有威力, 但是这些同伦不变量的计算通常都是很困难的. 然而, 利用幂零流形和可解流形之上的纤维

---

① 摘自《中国科学》2008 年第 38 卷, 第 1 期: 21-30.

结构,Heath 和 Keppelmann[①②] 成功地证明了在某种条件下 Nielsen 数和两个 Nielsen 型数是彼此相关的.

当 f 是三维幂零流形的自映射时,他们在本节中给出了 f 所有迭代的 Nielsen 型数 $NP_n(f)$ 和 $N\Phi_n(f)$ 的完整计算(见定理 2).为此,他们在同伦的意义下,分类了每一个三维幂零流形上的所有自映射.最一般的情形已被 Heath 和 Keppelmann 在文章 *Fibre techniques in Nielsen periodic point theory on nil and solvmanifolds* 和 *Fibre techniques in Nielsen periodic point theory on nil and solvmanifolds I* 中讨论,剩余部分在本文中加以研究.

动力系统理论中一个自然的问题是最小周期恰为 n 的周期点的存在性.同伦地考虑,一个称之为同伦最小周期集的新概念

$$\mathrm{HPer}(f) =: \bigcap_{g \simeq f} \{m \mid g^m(x) = x, g^q(x) \neq x,$$
$$q < m, \text{对某一} \ x \in X\}$$

被 Alsedà 等[③]引进.由于同伦最小周期集是在流形 X 上自映射 f 的小扰动下保持不变的,可以说 f 的同伦最小周期集 $\mathrm{HPer}(f)$ 描述了动力系统 f 的刚性部分.Jezierski 和 Marzantowicz[④] 给出了三维幂零流形上所有自映射的同伦最小周期集的特征,他们利用了这样的事实:每一个三维幂零流形 M 都有以 S^1 为纤维、T^2 为底的纤维结构,并且 M 上每一个自映射都同伦于一个保纤维的映射.从而,以前关于

————————————

① HEATH P R,KEPPELMANN E. Fibre techniques in Nielsen periodic point theory on nil and solvmanifolds. R. Mat. Rep. Acad. Sci. Canad.,16(6): 229-234(1994).

② HEATH P R,KEPPELMANN E. Fibre techniques in Nielsen periodic point theory on nil and solvmanifolds I. Topology Appl.,76(3):217-247(1997).

③ ALSEDÀ L,BALDWIN S,LLIBRE J,et al. Minimal sets of periods for torus maps via Nielsen numbers. Pacific J. Math,169(1):1-32(1995).

④ JEZIERSKI J,MARZANTOWICZ W. Homotopy minimal periods for maps of three-dimensional nilmanifolds. Pacific J. Math.,209(1):85-101(2003).

S^1 和 T^2 的结果可以被利用. 可是, 我们发现了他们工作中的一个错误. 本文的第二个目的就是改正这一错误, 并给出不同胚于三维环面的三维幂零流形上所有自映射的同伦最小周期集(见定理 3). 与文章 *Homotopy minimal periods for maps of three-dimensional nilmanifolds* 中使用的方法相比, 我们用来计算同伦最小周期集的方法要简单得多, 而且直接.

我们在文中写出了 Heisenberg 群中格点上同态的一般形式, 这可在文章 *Homotopy minimal periods for maps of three-dimensional nilmanifolds* 中找到; 讨论了三维幂零流形自映射 f 的线性化, 从而利用 Anosov 定理了解 f 的迭代的 Nielsen 数. 任意幂零流形以及 NR 型可解流形上自映射的线性化已经被人们考虑并论述过, 如在文章 *Homotopy methods in topological fixed and periodic points theory* 和 *Linearizations for maps of nilmanifolds and solvmanifolds* 中. 依靠这些准备性的内容, 对于三维幂零流形上的自映射, 我们确定了 Nielsen 型数和同伦最小周期集.

§2　三维幂零流形上的映射

设 G 是一个三维 Heisenberg 群, 即

$$G = \left\{ \begin{bmatrix} 1 & y & z \\ 0 & 1 & x \\ 0 & 0 & 1 \end{bmatrix} \middle| x, y, z \in \mathbb{R} \right\}$$

这样的 G 是唯一三维的、非交换的、连通且单连通的幂零 Lie 群, 参见文章 *Left invariant metrics and curvatures on connected, simply connected three-dimensional Lie groups*. 对于整数 $k > 0$, 我们考虑 G 的子群

$$\Gamma_k = \left\{ \begin{bmatrix} 1 & n & \dfrac{l}{k} \\ 0 & 1 & m \\ 0 & 0 & 1 \end{bmatrix} \middle| m,n,l \in \mathbb{Z} \right\}$$

这些都是 G 的一致格,而 G 的每个一致格都同构于某一 Γ_k.
设

$$x = \begin{bmatrix} 1 & 0 & 0 \\ 0 & 1 & 1 \\ 0 & 0 & 1 \end{bmatrix}$$

$$y = \begin{bmatrix} 1 & 1 & 0 \\ 0 & 1 & 0 \\ 0 & 0 & 1 \end{bmatrix}$$

$$z = \begin{bmatrix} 1 & 0 & -\dfrac{1}{k} \\ 0 & 1 & 0 \\ 0 & 0 & 1 \end{bmatrix}$$

我们得到 Γ_k 的表示如下

$$\Gamma_k = \langle x,y,z \mid [x,z]=[y,z]=1, [x,y]=z^k \rangle$$

在 Γ_k 中的每个元素都可唯一地写成形式

$$x^m y^n z^l = \begin{bmatrix} 1 & n & -\dfrac{l}{k} \\ 0 & 1 & m \\ 0 & 0 & 1 \end{bmatrix}$$

引理 1[①] G 的格上的任意同态 $\varphi : \Gamma_k \to \Gamma_k$ 都可由如下形式给出

$$\varphi(x) = x^\alpha y^\gamma z^\mu, \varphi(y) = x^\beta y^\delta z^\nu, \varphi(z) = z^{\alpha\delta - \beta\gamma} \qquad (1)$$

其中 $\alpha, \beta, \gamma, \delta, \mu, \nu \in \mathbb{Z}$. 等价地,$\varphi$ 的对应是

---

① 参见 *Homotopy minimal periods for maps of three-dimensional nilmanifolds* 中的命题 2.12.

$$\begin{bmatrix} 1 & q & -\dfrac{r}{k} \\ 0 & 1 & p \\ 0 & 0 & 1 \end{bmatrix} = x^p y^q z^r \longmapsto \begin{bmatrix} 1 & \gamma p + \delta q & -\dfrac{r'}{k} \\ 0 & 1 & \alpha p + \beta q \\ 0 & 0 & 1 \end{bmatrix}$$

$$= x^{\alpha p + \beta q} y^{\gamma p + \delta q} z^{r'}$$

其中

$$r' = (\alpha\delta - \beta\gamma)r - \alpha\gamma \frac{p(p-1)}{2}k -$$

$$\beta\delta \frac{q(q-1)}{2}k - \beta\gamma pqk + \mu p + \nu q$$

证明　注意到 Γ_k 的中心是 $\langle z \rangle$，则有 $\varphi(z) = z^\zeta$ 对某一 $\zeta \in \mathbb{Z}$ 成立. 记 $\varphi(x) = x^\alpha y^\gamma z^\mu$ 和 $\varphi(y) = x^\beta y^\delta z^\nu$，其中 $\alpha, \beta, \gamma, \delta, \mu, \nu \in \mathbb{Z}$. 由于 φ 保持关系 $[x, y] = z^k$，故必有

$$[\varphi(x), \varphi(y)] = \varphi(z)^k$$

即

$$(x^\alpha y^\gamma)(x^\beta y^\delta)(x^\alpha y^\gamma)^{-1}(x^\beta y^\delta)^{-1} = z^{k\zeta}$$

由此得到 $z^{k(\alpha\delta - \beta\gamma)} = z^{k\zeta}$.

设 $\Gamma \backslash G$ 是一个幂零流形. 因为 Γ 同构于某一 Γ_k，由 Mal'cev 的定理[①]，这个同构可唯一地扩张成 G 的自同构，并导出了从 $\Gamma \backslash G$ 到 $\Gamma_k \backslash G$ 的微分同胚. 于是可以假设给定的幂零流形是 $\Gamma_k \backslash G$.

由引理 1，任意一个同态 $\varphi : \Gamma_k \to \Gamma_k$ 都由整数集 $\{\alpha, \beta, \gamma, \delta, \mu, \nu\}$ 来确定. 记式 (1) 中的同态为 $\varphi^{\Gamma_k}_{\alpha, \beta, \gamma, \delta, \mu, \nu}$. 由 Mal'cev 的定理，每一同态 $\varphi^{\Gamma_k}_{\alpha, \beta, \gamma, \delta, \mu, \nu} : \Gamma_k \to \Gamma_k$ 可唯一地扩张成 G 上的 Lie 群同态，然后这一同态导出幂零流形 $\Gamma_k \backslash G$ 上的映射，仍记为 $\varphi^{\Gamma_k}_{\alpha, \beta, \gamma, \delta, \mu, \nu}$.

我们已能叙述三维幂零流形上自映射的分类定理.

定理 1　每一 $\Gamma_k \backslash G$ 上的自映射都同伦于一个形如

---

① MAL'CEV A I. On a class of homogeneous spaces. Trans. Amer. Math. Soc., 39:1-33(1951).

$\varphi_{\alpha;\beta,\gamma,\delta,\mu,\nu}^{\Gamma_k}:\Gamma_k\backslash G\to\Gamma_k\backslash G$ 的映射. 两个 $\Gamma_k\backslash G$ 上的映射 $\varphi_{\alpha;\beta,\gamma,\delta,\mu,\nu}^{\Gamma_k}$ 和 $\varphi_{\alpha',\beta',\gamma',\delta',\mu',\nu'}^{\Gamma_k}$ 相互同伦当且仅当 $\alpha=\alpha',\beta=\beta',\gamma=\gamma',\delta=\delta'$,并且

$$\begin{bmatrix}\mu & \nu\end{bmatrix}-\begin{bmatrix}\mu' & \nu'\end{bmatrix}\in k\cdot\mathrm{Im}\left(\begin{bmatrix}\alpha & \beta\\\gamma & \delta\end{bmatrix}\right)$$

证明　设 $f:\Gamma_k\backslash G\to\Gamma_k\backslash G$ 是幂零流形 $\Gamma_k\backslash G$ 上的自映射. 将 Γ_k 视为 $\Gamma_k\backslash G$ 的基本群,f 诱导一个在 Γ_k 上的同态 f_π. 由引理 1,$f_\pi=\varphi_{\alpha;\beta,\gamma,\delta,\mu,\nu}^{\Gamma_k}$,其中 $\alpha,\beta,\gamma,\delta,\mu,\nu$ 是整数. 由于幂零流形 $\Gamma_k\backslash G$ 是 $K(\Gamma_k,1)$ 型流形,因此有 $f\simeq\varphi_{\alpha;\beta,\gamma,\delta,\mu,\nu}^{\Gamma_k}$.

　　若

$$\begin{bmatrix}\mu & \nu\end{bmatrix}-\begin{bmatrix}\mu' & \nu'\end{bmatrix}\in k\cdot\mathrm{Im}\left(\begin{bmatrix}\alpha & \beta\\\gamma & \delta\end{bmatrix}\right)$$

则存在整数 m 和 n,使得

$$\begin{bmatrix}\mu & \nu\end{bmatrix}-\begin{bmatrix}\mu' & \nu'\end{bmatrix}=k\begin{bmatrix}n & -m\end{bmatrix}\begin{bmatrix}\alpha & \beta\\\gamma & \delta\end{bmatrix}$$

通过一个简单的计算得知 $w\varphi_{\alpha;\beta,\gamma,\delta,\mu,\nu}^{\Gamma_k}(u)w^{-1}=\varphi_{\alpha;\beta,\gamma,\delta,\mu',\nu'}^{\Gamma_k}(u)$ 对所有的 $u\in\Gamma_k$ 成立,其中

$$w=\begin{bmatrix}1 & n & -\dfrac{l}{k}\\0 & 1 & m\\0 & 0 & 1\end{bmatrix}\in\Gamma_k$$

由于幂零流形 $\Gamma_k\backslash G$ 是 $K(\Gamma_k,1)$ 型流形,可得出 $\varphi_{\alpha;\beta,\gamma,\delta,\mu,\nu}^{\Gamma_k}$ 和 $\varphi_{\alpha;\beta,\gamma,\delta,\mu',\nu'}^{\Gamma_k}$ 在 $\Gamma_k\backslash G$ 中同伦.

　　第二个结论的另一方面的证明只是一个简单的计算.

§3　Nielsen 数 $N(f^n)$

以 $\mathfrak{G}$ 记唯一的三维的、非交换的、连通且单连通的幂零 Lie 群的 Lie 代数,即

$$\mathfrak{S} = \left\{ \begin{bmatrix} 0 & b & c \\ 0 & 0 & a \\ 0 & 0 & 0 \end{bmatrix} \middle| a,b,c \in \mathbb{R} \right\}$$

已知指数映射 exp:$\mathfrak{S} \to G$ 是微分同胚,记其逆为 log,则有如下性质:

(1) 对于 Lie 群 G 上的任何同态 $\phi:G \to G$,有 Lie 代数 $\mathfrak{S}$ 上唯一的同态 $d\phi:\mathfrak{S} \to \mathfrak{S}$($\phi$ 的微分),使其如图 6 进行交换.

$$
\begin{array}{ccc}
G & \xrightarrow{\phi} & G \\
\log \downarrow & & \downarrow \log \\
\mathfrak{S} & \xrightarrow{d\phi} & \mathfrak{S}
\end{array}
$$

图 6

(2) 反之,对任何 Lie 代数同态 $d\phi:\mathfrak{S} \to \mathfrak{S}$,有 Lie 群上唯一的同态 $\phi:G \to G$,使其如图 6 进行交换.

事实上,指数映射是按如下对应给出的,即

$$\exp: \begin{bmatrix} 0 & b & c \\ 0 & 0 & a \\ 0 & 0 & 0 \end{bmatrix} \in \mathfrak{S} \mapsto \begin{bmatrix} 1 & b & c+\dfrac{ab}{2} \\ 0 & 1 & a \\ 0 & 0 & 1 \end{bmatrix} \in G$$

设

$$\boldsymbol{e}_1 = \begin{bmatrix} 0 & 0 & 0 \\ 0 & 0 & 1 \\ 0 & 0 & 0 \end{bmatrix}, \boldsymbol{e}_2 = \begin{bmatrix} 0 & 1 & 0 \\ 0 & 0 & 0 \\ 0 & 0 & 0 \end{bmatrix}, \boldsymbol{e}_3 = \begin{bmatrix} 0 & 0 & -\dfrac{1}{k} \\ 0 & 0 & 0 \\ 0 & 0 & 0 \end{bmatrix}$$

则有 $\exp(\boldsymbol{e}_1)=x, \exp(\boldsymbol{e}_2)=y, \exp(\boldsymbol{e}_3)=z$. 考虑在引理 1 中给出的在 Γ_k 中、进而在 G 中的同态 φ,我们可以看出它的微分 $d\varphi$ 如下

$$d\varphi(\boldsymbol{e}_1) = \alpha\boldsymbol{e}_1 + \gamma\boldsymbol{e}_2 + \left(\mu + \frac{\alpha\gamma}{2}k\right)\boldsymbol{e}_3$$

$$d\varphi(\boldsymbol{e}_2) = \beta\boldsymbol{e}_1 + \delta\boldsymbol{e}_2 + \left(\nu + \frac{\beta\delta}{2}k\right)\boldsymbol{e}_3$$

$$\mathrm{d}\varphi(\boldsymbol{e}_3) = (\alpha\delta - \beta\gamma)\boldsymbol{e}_3$$

因此,依照有序基$\{\boldsymbol{e}_1,\boldsymbol{e}_2,\boldsymbol{e}_3\}$,则

$$\mathrm{d}\varphi = \begin{bmatrix} \alpha & \beta & 0 \\ \gamma & \delta & 0 \\ \mu + \dfrac{\alpha\gamma}{2}k & \nu + \dfrac{\beta\delta}{2}k & \alpha\delta - \beta\gamma \end{bmatrix}$$

它的特征多项式是

$$\chi_\varphi(s) = (s - \det \boldsymbol{A}_\varphi)(s^2 - \mathrm{tr}\, \boldsymbol{A}_\varphi s + \det \boldsymbol{A}_\varphi)$$

其中

$$\boldsymbol{A}_\varphi = \begin{bmatrix} \alpha & \beta \\ \gamma & \delta \end{bmatrix}$$

注记 1　以上论证不仅对三维幂零流形,而且对所有维数的幂零流形上的映射都成立,证明可见文章 *Anosov theorem for coincidences on nilmanifolds* 中的定理 2.4.

定义　设 $f:\Gamma_k\backslash G \to \Gamma_k\backslash G$ 是由在式(1)中同态 φ 定义的连续映射. $\boldsymbol{A}_\varphi = \begin{bmatrix} \alpha & \beta \\ \gamma & \delta \end{bmatrix}$ 的特征值是 λ_1 和 λ_2,即

$$\lambda_i = \frac{(\alpha+\delta) \pm \sqrt{(\alpha-\delta)^2 + 4\beta\gamma}}{2}$$

称为 f 的基本特征值.

推论 1　设 $f:\Gamma_k\backslash G \to \Gamma_k\backslash G$ 是以 λ_1 和 λ_2 为基本特征值的映射,则 $f_*:\mathfrak{S} \to \mathfrak{S}$ 的特征多项式的特征值是 λ_1 和 λ_2 以及 $\lambda_1\lambda_2$.

注记 2　注意到幂零流形 $\Gamma_k\backslash G$ 有主纤维丛结构

$$Z(\Gamma_k)\backslash Z(G) \to \Gamma_k\backslash G \to (\Gamma/Z(\Gamma_k))\backslash(G/Z(G))$$

其中纤维 $Z(\Gamma_k)\backslash Z(G)$ 是圆周而基空间 $(\Gamma/Z(\Gamma_k))\backslash(G/Z(G))$ 是二维环面. 映射 $f = \varphi_{\alpha,\beta,\gamma,\delta,\mu,\nu}^{\Gamma_k}$ 是纤维映射,它诱导了基空间上的映射 $\bar{f}$ 和纤维上的映射 $\hat{f}$. 进一步,$\bar{f}$ 是由 $G/Z(G) = \mathbb{R}^2$ 上的线性映射 $\boldsymbol{A}_\varphi$ 所诱导的,而 $\hat{f}$ 是由 $Z(\Gamma_k)\backslash Z(G) = \mathbb{R}$ 上系数为 $\alpha\delta - \beta\gamma$ 的线性映射所诱导的. 于是

$$\deg(\bar{f}) = \det(\boldsymbol{A}_\varphi) = \lambda_1 \lambda_2 = \deg(\hat{f})$$

亦见文章 *Homotopy minimal periods for maps of three-dimensional nilmanifolds*.

由 Anosov 定理[①],映射 f 的 n 次迭代的 Lefschetz 数和 Nielsen 数由下式给出

$$L(f^n) = \det(I - d\varphi^n) = (1 - \det \boldsymbol{A}_\varphi^n)(1 - \operatorname{tr} \boldsymbol{A}_\varphi^n + \det \boldsymbol{A}_\varphi^n)$$
$$N(f^n) = |\det(I - d\varphi^n)| = |1 - \det \boldsymbol{A}_\varphi^n||1 - \operatorname{tr} \boldsymbol{A}_\varphi^n + \det \boldsymbol{A}_\varphi^n|$$

因此,我们有:

性质 1　设 $f : \Gamma_k \backslash G \to \Gamma_k \backslash G$ 是以 λ_1 和 λ_2 为基本特征值的连续映射,则

$$L(f^n) = (1 - \lambda_1^n \lambda_2^n)(1 - \lambda_1^n)(1 - \lambda_2^n)$$
$$N(f^n) = |1 - \lambda_1^n \lambda_2^n||1 - \lambda_1^n||1 - \lambda_2^n|$$

这意味着,对于三维幂零流形的连续映射,Lefschetz 数和 Nielsen 数由该映射的两个基本特征值完全决定. 在以下各节中,我们将证明这两个基本特征值也决定了各种 Nielsen 型数,如:$NP_n(f)$,$N\Phi_n(f)$ 和 $\mathrm{HPer}(f)$.

§4　Nielsen 型数 $NP_n(f)$ 和 $N\Phi_n(f)$

Heath 和 Keppelmann 在文章 *Fibre techniques in Nielsen periodic point theory on nil and solvmanifolds* 和 *Fibre techniques in Nielsen periodic point theory on nil and solvmanifolds I* 中证明了对于幂零流形的大多数映射,两个 Nielsen 型数 $NP_n(f)$ 和 $N\Phi_n(f)$ 能由 Nielsen 数计算得出.

① ANOSOV D V. The Nielsen numbers of maps of nil-manifolds. Uspehi Mat. Nauk,40(4):133-134(1985);Russian Math Survey,40:149-150(1985).

性质2[①] 设 f 是幂零流形上的映射. 如果 $N(f^n) \neq 0$, 那么对所有的 $m \mid n$, 有

$$N\Phi_m(f) = N(f^m), NP_m(f) = \sum_{q \mid m} \mu(q) N(f^{\frac{m}{q}})$$

其中 μ 是 Möbius 函数.

既然已经有了三维幂零流形上连续映射的 Nielsen 数的公式(性质1), 那么 $NP_n(f)$ 和 $N\Phi_n(f)$ 在 $N(f^n) \neq 0$ 时就是已知的了. 这里集中处理 $N(f^n) = 0$ 的情形.

下述引理给出了 $N(f^n) = 0$ 的充分必要条件.

引理2 设 $f: \Gamma_k \backslash G \to \Gamma_k \backslash G$ 是三维幂零流形 $\Gamma_k \backslash G$ 上以 λ_1 和 λ_2 为基本特征值的映射, 则 $N(f^n) = 0$ 当且仅当如下条件之一成立:

(1) $(\lambda_1\lambda_2, \lambda_1 + \lambda_2) = (1, l)$;

(2) $(\lambda_1\lambda_2, \lambda_1 + \lambda_2) = (-1, l)$, 并且 n 是偶数;

(3) $(\lambda_1\lambda_2, \lambda_1 + \lambda_2) = (l, l+1)$;

(4) $(\lambda_1\lambda_2, \lambda_1 + \lambda_2) = (l, -l-1)$, 并且 n 是偶数.

证明 回顾 $N(f^n) = |1 - \lambda_1^n \lambda_2^n| |1 - \lambda_1^n| |1 - \lambda_2^n|$. 如果 $N(f^n) = 0$, 那么 $|1 - \lambda_1^n \lambda_2^n| = 0$, 或者 $|1 - \lambda_1^n| \cdot |1 - \lambda_2^n| = 0$.

假定 $|1 - \lambda_1^n \lambda_2^n| = 0$. 因为 $\lambda_1\lambda_2$ 是一个整数, 有两种可能: (i) $\lambda_1\lambda_2 = 1$, (ii) $\lambda_1\lambda_2 = -1$, 并且 n 是偶数. 而这些恰好是本引理所述的前两种情形, 其中 l 是整数.

假定 $|1 - \lambda_1^n| |1 - \lambda_2^n| = 0$. 当 λ_1 和 λ_2 不是实数时, 它们是共轭的. 条件 $|1 - \lambda_1^n| |1 - \lambda_2^n| = 0$ 蕴含 $|\lambda_1| = |\lambda_2| = 1$. 由于它们的实部 $\mathrm{Re}\, \lambda_1 = \mathrm{Re}\, \lambda_2 = \frac{1}{2}(\lambda_1 + \lambda_2)$ 是半整数, 所以两个基本特征值必须是如下偶对之一: $(\mathrm{i}, -\mathrm{i})$,

---

① 参见 *Fibre techniques in Nielsen periodic point theory on nil and solvmanifolds* 中的定理 1 及 *Fibre techniques in Nielsen periodic point theory on nil and solvmanifolds I* 中的定理 1.2.

$\left(\dfrac{1}{2}+\dfrac{\sqrt{3}}{2}i,\dfrac{1}{2}-\dfrac{\sqrt{3}}{2}i\right)$ 或 $\left(-\dfrac{1}{2}+\dfrac{\sqrt{3}}{2}i,-\dfrac{1}{2}-\dfrac{\sqrt{3}}{2}i\right)$. 相应地，$(\lambda_1\lambda_2,\lambda_1+\lambda_2)=(1,0),(1,1)$ 或 $(1,-1)$，这些都含于(1)中.

当 λ_1 和 λ_2 是实数时，条件 $|1-\lambda_1^n||1-\lambda_2^n|=0$ 蕴含着 $\lambda_1^n=1$ 或 $\lambda_2^n=1$. 这些含于(3)和(4)中，其中 l 和 $-l$ 分别是另外的特征值，自然是整数.

反之，很容易验证以上 4 个条件的每一个都保证了 $N(f^n)=0$.

事实上，这 4 种情形都有可能发生. 例如：A_φ 可以分别选成

$$\begin{bmatrix} l & 1 \\ -1 & 0 \end{bmatrix},\begin{bmatrix} l & 1 \\ 1 & 0 \end{bmatrix},\begin{bmatrix} l & 0 \\ 0 & 1 \end{bmatrix},\begin{bmatrix} -l & 0 \\ 0 & -1 \end{bmatrix}$$

我们即可得到：

推论 2 设 $f:\Gamma_k\backslash G\to\Gamma_k\backslash G$ 是一个映射，那么：

(1) 当 $N(f^n)\neq 0$ 时，则对所有的 $m\mid n$ 有 $N(f^m)\neq 0$；

(2) 当对某一奇数 n 有 $N(f^n)=0$ 时，则对所有的 m 有 $N(f^m)=0$；

(3) 当对某一偶数 n 有 $N(f^n)=0$ 时，则对所有的偶数 m 有 $N(f^m)=0$.

由基本特征值的定义，我们有：

推论 3 设 $f:\Gamma_k\backslash G\to\Gamma_k\backslash G$ 是一个由式(1)中的同态 φ 确定的连续映射，则 $N(f^n)=0$ 当且仅当 $\alpha,\beta,\gamma,\delta$ 满足下列条件之一：

(1) $\alpha\delta-\beta\gamma=1$；

(2) $\alpha\delta-\beta\gamma=-1$，并且 n 是偶数；

(3) $\beta\gamma=(\alpha-1)(\delta-1)$；

(4) $\beta\gamma=(\alpha+1)(\delta+1)$，并且 n 是偶数.

对比引理 2 和性质 1，我们得到：

推论 4 设 $f:\Gamma_k\backslash G\to\Gamma_k\backslash G$ 是以 λ_1 和 λ_2 为基本特征值的映射，则如下陈述等价：

(1) $L(f) = 0$；

(2) $N(f) = 0$；

(3) $(\lambda_1 \lambda_2, \lambda_1 + \lambda_2) = (1, l)$ 或 $(l, l+1)$；

(4) 对所有的 k 有 $N(f^k) = 0$；

(5) 对所有的 k 有 $N\Phi_k(f) = 0$。

现在让我们考虑当 $N(f^n) = 0$ 时的 Nielsen 型数 $N\Phi_n(f)$。

性质 3 设 $f : \Gamma_k \backslash G \to \Gamma_k \backslash G$ 是一个以 λ_1 和 λ_2 为基本特征值的映射。假设 $N(f^n) = 0$ 并且 $n = 2^r n_0$，其中 r 是非负整数，n_0 是奇数，则

$$N\Phi_n(f) = \begin{cases} 2 \mid 1 - \lambda_1^{n_0} \mid \mid 1 - \lambda_2^{n_0} \mid, & \text{当} (\lambda_1 \lambda_2, \lambda_1 + \lambda_2) = (-1, l), r > 0 \\ 2 \mid 1 - l^{2 n_0} \mid, & \text{当} (\lambda_1 \lambda_2, \lambda_1 + \lambda_2) = (l, -l-1), r > 0 \\ 0, & \text{其他} \end{cases}$$

证明 由于 $N(f^n) = 0$，偶对 $(\lambda_1 \lambda_2, \lambda_1 + \lambda_2)$ 有引理 2 列出的 4 种可能性。

当 $(\lambda_1 \lambda_2, \lambda_1 + \lambda_2) = (1, l)$ 或者 $(\lambda_1 \lambda_2, \lambda_1 + \lambda_2) = (l, l+1)$ 时，则对所有的 m 有 $N(f^m) = 0$。该映射没有任何周期的本质周期轨，于是 $N\Phi_n(f) = 0$。

假设 $(\lambda_1 \lambda_2, \lambda_1 + \lambda_2) = (-1, l)$ 并且 n 是偶数。当 $l = \pm 1$，即 $(\lambda_1 \lambda_2, \lambda_1 + \lambda_2) = (1, -2)$ 或 $(-1, 0)$ 时，那么对所有的 m 有 $N(f^m) = 0$。这样我们的结论成立。此外，当 $l \neq \pm 1$ 时，$N(f^m) \neq 0$ 对所有奇数成立。特别地，$N(f^{n_0}) \neq 0$。满足 $q \mid n$ 的周期为 q 的本质周期轨道类等同于满足 $q \mid n_0$ 的周期为 q 的本质周期轨道类。由定义可知 $N\Phi_n(f) = N\Phi_{n_0}(f)$。注意到 $\lambda_1 \lambda_2 = -1$。由性质 1 和性质 2，有

$$N\Phi_n(f) = N\Phi_{n_0}(f) = 2 \mid 1 - \lambda_1^{n_0} \mid \mid 1 - \lambda_2^{n_0} \mid$$

在 $(\lambda_1 \lambda_2, \lambda_1 + \lambda_2) = (l, -l-1)$ 并且 n 为偶数的情形，我们类似地得到 $N\Phi_n(f) = N\Phi_{n_0}(f) = 2 \mid 1 - l^{2n_0} \mid$。此处，基本特征值是 -1 和 $-l$。

如果 $N(f^n) = 0$，那么由定义知 $NP_n(f) = 0$。但是，$N\Phi_n(f)$ 可以是非零的。由上面的性质，这一现象的确发生

在三维幂零流形的某些映射上,这时 n 是偶数,并且或者 $(\lambda_1\lambda_2,\lambda_1+\lambda_2)=(-1,l)$ 及 $l\neq 0$,或者 $(\lambda_1\lambda_2,\lambda_1+\lambda_2)=(l,-l-1)$ 及 $l\neq\pm 1$. 由引理 2,这些条件等价于:对偶数 m 有 $N(f^m)=0$ 并且对奇数 m 有 $N(f^m)\neq 0$.

现在,我们将三维幂零流形 $\Gamma_k\backslash G$ 上 Nielsen 型数的计算总结如下:

定理 2 设 $f:\Gamma_k\backslash G\rightarrow\Gamma_k\backslash G$ 是连续映射,则:

(1) 当 $N(f^n)=0$ 时,有 $NP_n(f)=0$ 和 $N\Phi_n(f)=N(f^{n_0})$,其中 $n=2^r n_0$,且 $r\geqslant 0$ 并且 n_0 是奇数.

(2) 当 $N(f^n)\neq 0$ 时,有 $NP_n(f)=\sum_{q|n}\mu(q)N(f^{\frac{n}{q}})$ 和 $N\Phi_n(f)=N(f^n)$,其中 μ 是 Möbius 函数.

§5 同伦最小周期集 HPer(f)

下面的引理源于姜和 Llibre 在文章 *Minimal sets of periods for torus maps*(定理 3.4)中的结果.

引理 3 设 $f:M\rightarrow M$ 是在维数不小于 3 的光滑流形上的一个连续映射.若对所有的 n,或者 $N(f^n)=\sum_{m|n}NP_m(f)$ 或者 $N(f)=0$,当有一个整数矩阵 $\boldsymbol{A}$ 使得 $N(f^n)=|\det(\boldsymbol{I}-\boldsymbol{A}^n)|$ 对所有 n 成立时,则

$$\text{HPer}(f)=\{n\mid NP_n(f)\neq 0\}$$
$$=\{n\mid N(f^n)\neq 0,N(f^n)\neq N(f^{\frac{n}{q}})$$
$$\text{对所有素数 } q\mid n,q\leqslant n \text{ 成立}.\}$$

证明 由于 $NP_n(f)$ 是最小周期为 n 的周期点数的一个下界,由定义,我们有 $\text{HPer}(f)\supset\{n\mid NP_n(f)\neq 0\}$. 如果 $NP_n(f)=0$,那么由文章 *Wecken theorem for fixed and periodic points*,存在一个同伦于 f 的映射 g,它没有最小周期为 n 的周期点.我们完成了第 1 个等式.

另一个等式的证明与文章 *Minimal sets of periods for torus maps* 中定理 3.4 的证明完全相同(参见文章

Homotopy minimal periods).

我们即可得

$$1 \notin \mathrm{HPer}(f) \Leftrightarrow N(f) = 0$$

由推论 4,这蕴含着对于所有的 n,有 $N(f^n) = 0$,从而 $n \notin \mathrm{HPer}(f)$,这说明 $\mathrm{HPer}(f) = \varnothing$.

引理 4　设 $f: \Gamma_k \backslash G \to \Gamma_k \backslash G$ 是以 λ_1 和 λ_2 为基本特征值的映射,则如下陈述等价:

(1) $1 \notin \mathrm{HPer}(f)$;

(2) $(\lambda_1 \lambda_2, \lambda_1 + \lambda_2)$ 有形式 $(1, l)$ 或 $(l, l+1)$,其中 l 是整数;

(3) $N(f) = 0$;

(4) $\mathrm{HPer}(f) = \varnothing$.

下面的引理刻画了满足条件 $1 \in \mathrm{HPer}(f)$ 和 $2 \notin \mathrm{HPer}(f)$ 的连续映射 f.

引理 5　设 $f: \Gamma_k \backslash G \to \Gamma_k \backslash G$ 是以 λ_1 和 λ_2 为基本特征值的映射. 假定 $1 \in \mathrm{HPer}(f)$,则 $2 \notin \mathrm{HPer}(f)$,当且仅当 $(\lambda_1 \lambda_2, \lambda_1 + \lambda_2)$ 是 $(0, 0)$,$(0, -2)$,$(-2, 0)$,$(-2, 2)$,$(-1, l)$ 且 $l \neq 0$,或 $(l, -l-1)$ 且 $l \neq \pm 1$ 这几种情形之一.

证明　由引理 3,可看出 $2 \notin \mathrm{HPer}(f)$ 当且仅当 $N(f^2) = 0$ 或者 $N(f^2) = N(f) \neq 0$. 在后一种情形,我们有等式 $(1 + \lambda_1)(1 + \lambda_2) = \pm 1$,以及 $(1 + \lambda_1 \lambda_2) = \pm 1$. 这蕴含 $(\lambda_1 \lambda_2, \lambda_1 + \lambda_2) = (0, 0)$,$(0, -2)$,$(-2, 0)$,或 $(-2, 2)$

由条件 $N(f^2) = |1 - \lambda_1^2 \lambda_2^2||1 - \lambda_1^2||1 - \lambda_2^2| = 0$ 可推知 $\lambda_1 = \pm 1, \lambda_2 = \pm 1$ 或者 $\lambda_1 \lambda_2 = \pm 1$. 既然 $N(f) \neq 0$,我们有 $\lambda_1 = -1, \lambda_2 = -1$ 或 $\lambda_1 \lambda_2 = -1$. 如果 $\lambda_1 \lambda_2 = -1$,那么 $\lambda_2 = -1/\lambda_1$ 并且 $\lambda_1 + \lambda_2 = \dfrac{\lambda_1^2 - 1}{\lambda_1}$ 是某一整数 l. 但是,由 $\lambda_1 \neq 1$ 和 $\lambda_2 \neq 1$ 可知 $l \neq 0$. 在此情形下,我们得出

$$(\lambda_1 \lambda_2, \lambda_1 + \lambda_2) = (-1, l), l \neq 0$$

当某一 $\lambda_i = -1$ 时,则另一 λ_j 是一个整数,并且不是 ± 1. 从而得到

$$(\lambda_1\lambda_2,\lambda_1+\lambda_2)=(l,-l-1),l\neq\pm1$$

至此,引理得证.

引理 6 设 $f:\Gamma_k\backslash G\to\Gamma_k\backslash G$ 是以 λ_1 和 λ_2 为基本特征值的映射.设 m 是不小于3的一个整数,使得 $N(f^m)\neq0$.如果 $m\notin\mathrm{HPer}(f)$,那么 $(\lambda_1\lambda_2,\lambda_1+\lambda_2)=(0,0)$ 或 $(0,-1)$.

证明 既然 $N(f^m)\neq0$ 并且 $m\notin\mathrm{HPer}(f)$,由引理3,整数 m 将有一个素因子 p,使得 $N(f^m)=N(f^{\frac{m}{p}})$.这样,有

$$|1-\lambda_1^m\lambda_2^m||1-\lambda_1^m||1-\lambda_2^m|$$
$$=|1-\lambda_1^{\frac{m}{p}}\lambda_2^{\frac{m}{p}}||1-\lambda_1^{\frac{m}{p}}||1-\lambda_2^{\frac{m}{p}}|$$

于是

$$|1+(\lambda_1\lambda_2)^{\frac{m}{p}}+\cdots+(\lambda_1\lambda_2)^{\frac{(p-1)m}{p}}||1+$$
$$\lambda_1^{\frac{m}{p}}+\cdots+\lambda_1^{\frac{(p-1)m}{p}}||1+\lambda_2^{\frac{m}{p}}+\cdots+\lambda_2^{\frac{(p-1)m}{p}}|=1$$

由于 $\lambda_1\lambda_2$ 和 $\lambda_1+\lambda_2$ 都是整数,上面的等式等价于

$$1+(\lambda_1\lambda_2)^{\frac{m}{p}}+\cdots+(\lambda_1\lambda_2)^{\frac{(p-1)m}{p}}=\pm1$$

$$(1+\lambda_1^{\frac{m}{p}}+\cdots+\lambda_1^{\frac{(p-1)m}{p}})(1+\lambda_2^{\frac{m}{p}}+\cdots+\lambda_2^{\frac{(p-1)m}{p}})=\pm1$$

前一个等式也就是

$$\frac{1-(\lambda_1\lambda_2)^m}{1-(\lambda_1\lambda_2)^{\frac{m}{p}}}=\pm1$$

因为 $N(f^m)\neq0$,所以 $(\lambda_1\lambda_2)^m\neq1$.则必有 $\lambda_1\lambda_2=0$.再由第2个等式得出 $(\lambda_1\lambda_2,\lambda_1+\lambda_2)=(0,0)$ 或 $(0,-1)$.

定理 3 设 $f:\Gamma_k\backslash G\to\Gamma_k\backslash G$ 是以 λ_1 和 λ_2 为基本特征值的映射,则:

(1) $(\lambda_1\lambda_2,\lambda_1+\lambda_2)=(1,l)$ 或 $(l,l+1)$ 当且仅当 $\mathrm{HPer}(f)=\varnothing$;

(2) $(\lambda_1\lambda_2,\lambda_1+\lambda_2)=(0,0)$ 或 $(0,-1)$ 当且仅当 $\mathrm{HPer}(f)=\{1\}$;

(3) $(\lambda_1\lambda_2,\lambda_1+\lambda_2)=(0,-2),(-2,0)$ 或 $(-2,2)$ 当且仅当 $\mathrm{HPer}(f)=\mathbb{N}\backslash\{2\}$;

(4) $(\lambda_1\lambda_2,\lambda_1+\lambda_2)=(-1,l)$,其中 $l\neq0$,或者 $(\lambda_1\lambda_2,$

$\lambda_1 + \lambda_2) = (l, -l-1)$，其中 $l \neq 0, \pm 1$ 当且仅当 $\mathrm{HPer}(f) = \mathbb{N} \backslash 2\mathbb{N}$；

(5) 其他情形，$\mathrm{HPer}(f) = \mathbb{N}$.

证明 在这 5 种情形中，注意到集 $\mathrm{HPer}(f)$ 是互不相同的. 我们只需证"当".

情形(1)：由引理 4 可得.

情形(2)：如果 $(\lambda_1\lambda_2, \lambda_1 + \lambda_2) = (0, 0)$，两个基本特征值都是 0. 这时，$N(f^n) = 1$ 对所有 n 成立. 由引理 3 知，$\mathrm{HPer}(f) = \{1\}$. 如果 $(\lambda_1\lambda_2, \lambda_1 + \lambda_2) = (0, -1)$，两个基本特征值是 0 和 -1. 这时，$N(f^n) = 1 - (-1)^n$ 对所有 n 成立. 再由引理 3 知，$\mathrm{HPer}(f) = \{1\}$.

情形(3)：通过一个计算，得

$$N(f^n) = \begin{cases} |1-(-2)^n| \, |1-2^{\frac{n}{2}}| \, |1-(-1)^n 2^{\frac{n}{2}}|, & \text{当} (\lambda_1\lambda_2, \lambda_1 + \lambda_2) = (-2, 0) \\ |1-(-2)^n|, & \text{当} (\lambda_1\lambda_2, \lambda_1 + \lambda_2) = (0, -2) \\ |1-(-2)^n| \, |1-(1+\sqrt{3})^n| \, |1-(1-\sqrt{3})^n|, & \text{当} (\lambda_1\lambda_2, \lambda_1 + \lambda_2) = (-2, 2) \end{cases}$$

由于 $N(f^2) = N(f) \neq 0$，因此有 $1 \in \mathrm{HPer}(f)$ 并且 $2 \notin \mathrm{HPer}(f)$. 注意到 $N(f^n) \neq 0$ 对所有 $n \geqslant 3$ 成立. 由引理 6 知，$n \in \mathrm{HPer}(f)$ 对所有 $n \geqslant 3$ 成立. 于是，$\mathrm{HPer}(f) = \mathbb{N} \backslash \{2\}$.

情形(4)：由于对所有的偶数 n 有 $N(f^n) = 0$，所以 $\mathrm{HPer}(f) \subset \mathbb{N} \backslash 2\mathbb{N}$. 而由 $N(f) \neq 0$ 知 $1 \in \mathrm{HPer}(f)$. 对于任何一个大于 1 的奇数 m，仍有 $N(f^m) \neq 0$. 由于 $(\lambda_1\lambda_2, \lambda_1 + \lambda_2) \neq (0, 0)$ 或 $(0, -1)$，由引理 6 可知，$m \in \mathrm{HPer}(f)$. 于是，$\mathrm{HPer}(f) = \mathbb{N} \backslash 2\mathbb{N}$.

在剩下的情形，由引理 4，有 $N(f) \neq 0$. 这样，$1 \in \mathrm{HPer}(f)$. 引理 5 保证了 $2 \in \mathrm{HPer}(f)$. 因为对任何 m 有 $N(f^m) \neq 0$，由引理 6 得知，$\mathrm{HPer}(f)$ 包含所有不小于 3 的整数，所以 $\mathrm{HPer}(f) = \mathbb{N}$.

现将我们得出的 $\mathrm{HPer}(f)$ 列于表 1：

表 1

$(\lambda_1\lambda_2,\lambda_1+\lambda_2)$	HPer(f)
$(1,l)$	$\varnothing$
$(l,l+1)$	$\varnothing$
$(0,0),(0,-1)$	$\{1\}$
$(0,-2),(-2,0),(-2,2)$	$\mathbb{N}\backslash\{2\}$
$(-1,l)$ 其中 $l\neq0$	$\mathbb{N}\backslash2\mathbb{N}$
$(l,-l-1)$ 其中 $l\neq0,\pm1$	$\mathbb{N}\backslash2\mathbb{N}$
其他	$\mathbb{N}$

注记 表 1 中的前 4 个情形和第 6 个情形与文章 *Homotopy minimal periods for maps of three-dimensional nilmanifolds* 中的一致. 但是, 当 $(\lambda_1\lambda_2,\lambda_1+\lambda_2)=(-1,2)$ 时, 对应映射的同伦最小周期集在文章 *Homotopy minimal periods for maps of three-dimensional nilmanifolds* 中声称是 $\mathbb{N}$, 可实际上它应是 $\mathbb{N}\backslash2\mathbb{N}$.

据戚心茹在复旦大学网站的"复旦数学家"中的报道:

2021 年 7 月, 复旦大学上海数学中心的王国祯副教授在国际顶尖数学期刊《数学学报》(*Acta Mathematica*) 在线发表了他与合作者 B. Gheorghe、徐宙利的论文《稳定同伦范畴的母体形变的特殊纤维是代数的》(暂译名: *The special fiber of the motivic deformation of the stable homotopy category is algebraic*). 这是他五年以来在国际顶尖刊物发表的第五篇论文, 也是复旦大学上海数学中心成立以来第六篇登上数学四大顶尖期刊的文章. 在文章中, 他自己创造"工具", 探索无穷维的拓扑空间, 他在站上"巨人的肩膀"前进的过程中, 最终计算出数学界学者们"有生之年"的结果.

"数学四大期刊"是国际数学领域四大顶级刊物, 刊载数学研究领域的重要突破性成果.1998—2017年,中国数学研究者在四大刊发表98篇论文.截至2021年,复旦大学上海数学中心的陈佳源、沈维孝、王国祯、李骏、周杨已经在"数学四大期刊"上连续发表五篇文章,彰显了数学中心基础数学研究的深度和厚度.

一、剪开那条麦比乌斯带

我们可以想象一下,把一条麦比乌斯带从中间剪开,你将会得到什么模型呢?大部分人的想法都是沿着纸面的中线剪开,但是,如果你的想象力足够充分的话,换个角度,你还可以沿着纸片薄薄的侧面切开,那又会是什么情形呢?

这是王国祯留给学生的一道课后习题,让同学们发挥想象力,感受探索空间的奥秘.

这是一个有趣的数学问题,它不是为了证明什么,而是想看看不同的操作能够产生什么样有意思的结果.王国祯说:"当你的数学研究走得比较远时,找问题比做问题更重要."

对王国祯来说,他的研究工作也是在不断地寻找问题,解决问题.

假如把普通几何看作硬的乒乓球的话,那么拓扑几何空间就像篮球,是软的.很多普通情况下做不到的事,拥有弹性的拓扑空间就能够做到.换言之,能够自由变换的拓扑空间拥有更多的可能性.王国祯的工作,就是尽可能多地推算出这些存在的可能性.

从本科开始,王国祯就对代数拓扑产生兴趣.在学习了北京大学范后宏老师的同伦论课程后,他迷恋上了拓扑学独特的解题方式."很难的微分方程,基于拓扑的一些特性,可以不用考虑其他细节,就能得出直观的解答.我觉得这是一个很神奇的、非常有效的解决问题的方法."王国祯说.

代数拓扑带给他看问题的另一个角度.他开始尝试从侧面切开那条麦比乌斯带.就像鱼从小溪游入了大海,王国祯走进数学广袤无边的世界,在不断思考中前行.

二、创造出"工具"

王国祯所计算的球面同伦群,是数学研究的一个基本问题,也是代数拓扑中的重要问题之一.问题提出一个世纪以来,学界将计算推进到球面的第 59 个稳定同伦群.王国祯与合作者们突破了 60 维的难关,将计算推进到第 90 维、第 120 维,解决了广义庞加莱猜想在奇数维情形的最后一个问题.

在最新发布的论文中,他提出了母体(motivic)同伦范畴中的周 t —结构.利用该 t —结构,王国祯与合作者证明了复数域上的母体形变的特殊纤维是代数的,以及模 τ 的母体亚当斯(motivic Adams)谱序列与代数诺维科夫(Novikov)谱序列是同构的.该结果给出了计算球面稳定同伦群的一个全新的计算方法,使球面稳定同伦群的计算顺利向百维推进,是同伦论领域的重要突破.

一维、二维的球面尚属于常见的事物,但三维就已经变得复杂,进入了更高的维度.59,60,90,120,这些维度不是简单的数字,每一维度的计算方法本质上都是完全不同的,需要给出不同的解法.这意味着前人的研究成果可提供的借鉴是有限的,不打开新的思路,研究就无法推进.这个问题的难点在于缺少有效的工具去计算.

王国祯一开始希望能够写一个计算机程序,以实现简单的辅助计算.用传统 C 语言所编写的程序,在运行过程中会产生很多漏洞,导致实验一次又一次失败.彼时王国祯觉得这并不算什么,球面同伦群的计算成果有一些周期性的特质,往往几十年才能实现一次重大突破,学界关于此的研究已有近十年的停滞,挑战者不断失败,而他只不过是其中之一罢了.

对于未知的渴求,让他陷入沉思,无时无刻不在思考着问题的解决路径,天花板、水面、地面都成了他脑中推演时的草稿纸.偶一间隙获得灵感,王国祯便可以连续多日算个不停.

　　"有一天我突发奇想,觉得我是不是应该学习一种新的计算机语言?说不定可以解决我之前的问题."王国祯回忆道.

　　于是,他试着用F♯语言重新编写程序.F♯语言有着C语言无法比拟的优势,能够大大减少运行中的漏洞,实验终于成功了.但同时,普通的电脑承载不了越来越庞大的数据,需要性能更强劲的服务器才能继续运行.然而,当王国祯试图在高级计算机上运行F♯语言程序时,他又遇到了机器无法编译的问题.

　　最终,他只能将F♯语言程序重新翻译回C语言,他惊喜地发现,经过反复的试验,吸收了数次失败的经验,程序终于能够顺利地运行了.就这样,王国祯创造出了新的工具.

　　通过将计算机的运行结果与合作者阿萨克森(Isaksen)之前的计算结果做比对,王国祯与合作者发现,通过计算机生成的代数诺维科夫谱序列的数据与通过经典方法得到的模τ的母体亚当斯谱序列的数据完全一致.这是一个振奋人心的发现,表明了一大类复杂的亚当斯微分可以通过计算机完全机械地得到.这是几十年来拓扑学家梦寐以求的方法.王国祯与合作者进一步研究了母体同伦理论,提出来周t—结构的概念,从而在理论上证明了前面的实验结果.这就给出了一个计算球面稳定同伦群的一个全新的工具.

　　以此为起点,他与合作者们一起,提出了很多新的工具,在代数拓扑的核心问题上取得了突破.

　　"我的研究就是建立各种工具,希望能够应用在同伦群的研究上,从而揭示一些同伦群的规律."王国祯说.

三、课程"几何拓扑选讲",面向"数学英才试验班"的本科生

　　王国祯这学期开了一门"几何拓扑选讲"的课程,专门面向"数学英才试验班"的本科生,以培养学生的基础数学能力.走进课堂时,他穿着半旧的格子衬衫,只带了两本教

科书和一瓶水.

课上,他写下一段段板书,讲解理论的定义和推导过程,频频抛出衍生的课后习题供学生们课后思考,不多时,两块黑板便铺满笔迹.

同学们已经适应了王老师简洁而快速的授课风格,目光往返于黑板与笔记之间,一时间,教室内悄然无声.

"王老师的这门课很深奥,他上课经常说'我们可以想象一下',但他往往能讲得准确而有体系."2020级数学系本科生孙艺青说:"他也会说一些生动的例子来辅助我们思考,调节课堂氛围."

在讲到米田(Yoneda)引理时,为了帮助同学们理解事物之间的映射关系,王国祯这么说:"假设社会上有两个人,他们两人的社会关系是完全一样的,那么可以看作是同一个人,就像电视剧里面的双胞胎扮演了同一个人."

带着"培养出未来的数学家"的目标,数学英才班于2020年秋季开班,一年多以来,优秀的师资收获了学生们的好评.进度快、程度深、内容难,成了英才班课程的特点.

作为数学英才班的授课老师,王国祯道出了他对学生们的期许.他希望同学们能对数学产生兴趣,将来能在数学领域继续精进最好,就算转向其他方向,也能通过英才班课程的培养掌握好的基础.

"数学不仅能影响人的思维方式,也是一项基础技能."王国祯说.

四、删繁就简

毕加索早年的画复杂而写实.随着年龄的上升与画艺的精进,他不断做减法,让自己的画作最终只留下简单的几何形状.

这与王国祯的思想不谋而合.

代数拓扑深奥而精微,多年的潜心研究使他能够删繁就简,将目光锁定在研究对象必要的元素上.

"数学里有一个非常重要的思想,就是把不重要的数据、几何结构忘掉,忘掉的越多越能从崭新的视角去认识问题."王国祯说:"研究同伦问题也是一样,你需要忘掉研究对象的刚性结构,想象它是软的."

在王国祯未来的研究中,他计划解决存在特质的 126 维同伦群问题,将球面同伦群的问题继续深入.

摒弃无关要素、抓住核心问题,是数学研究的理想状态,这也是王国祯努力的方向.

五、应邀在 2022 年国际数学家大会上做报告

王国祯应邀在今年的国际数学家大会上做 45 分钟报告,他将分享名为《球面稳定同伦群与母体同伦论》(*Stable homotopy groups of spheres and motivic homotopy theory*)的报告,分享自己的球面同伦群研究成果.

国际数学家大会(International Congress of Mathematicians,简称 ICM),是由国际数学联盟(IMU)主办的国际数学界规模最大、最重要的会议,被誉为数学界的奥林匹克盛会.2022 年 7 月 6 日至 14 日,大会将于俄罗斯圣彼得堡举行.大会上,有约 200 位数学家受邀做学术报告,分享他们在各自领域中取得的成果与进展.这些报告代表国际数学界最高水平.

让我们恭喜他,期待他进一步的研究成果.

王国祯履历简介:

2004—2011 年在北京大学就读,获学士、硕士学位.

2015 年获得麻省理工学院博士学位.

2015—2016 年在哥本哈根大学从事博士后研究.

2016—2018 年在上海数学中心从事博士后研究.

2018 年,作为青年研究员正式加入上海数学中心.

2020 年,晋升为长聘副教授.

在当前的这个互联网阅读环境中,像本书这样艰深的学术著作是少有人问津的.

这是互联网原住民们所面对的现实:住在一个充满陌生人的拥挤社区,一个破碎而断裂的世界.人们看到的信息被多轮加工,被切

割,缺乏上下文的语境.

媒体文化研究者波兹曼在专著《娱乐至死》中预言过电视时代的可能结局,"真正发生的是公众已经适应了没有连贯性的世界,并且已经被娱乐得麻木不仁了."这同样适用于如今的网络世界,一些事件被剥夺了与过去、未来或其他任何事件的关联 —— 连贯性消失了,自相矛盾存在的条件也随之消失了.在这个断裂的网络世界中,够语出惊人,就能获取注意成为"明星",娱乐、鄙夷、消费、羡慕之情可以并存.

我们所做的,就是对这种阅读环境的一种"反动".

刘培杰
2022. 11. 4
于哈工大

刘培杰数学工作室
已出版(即将出版)图书目录——原版影印

书　名	出版时间	定　价	编号
数学物理大百科全书. 第1卷(英文)	2016－01	418.00	508
数学物理大百科全书. 第2卷(英文)	2016－01	408.00	509
数学物理大百科全书. 第3卷(英文)	2016－01	396.00	510
数学物理大百科全书. 第4卷(英文)	2016－01	408.00	511
数学物理大百科全书. 第5卷(英文)	2016－01	368.00	512
zeta 函数,q-zeta 函数,相伴级数与积分(英文)	2015－08	88.00	513
微分形式:理论与练习(英文)	2015－08	58.00	514
离散与微分包含的逼近和优化(英文)	2015－08	58.00	515
艾伦·图灵:他的工作与影响(英文)	2016－01	98.00	560
测度理论概率导论,第2版(英文)	2016－01	88.00	561
带有潜在故障恢复系统的半马尔柯夫模型控制(英文)	2016－01	98.00	562
数学分析原理(英文)	2016－01	88.00	563
随机偏微分方程的有效动力学(英文)	2016－01	88.00	564
图的谱半径(英文)	2016－01	58.00	565
量子机器学习中数据挖掘的量子计算方法(英文)	2016－01	98.00	566
量子物理的非常规方法(英文)	2016－01	118.00	567
运输过程的统一—非局部理论:广义波尔兹曼物理动力学,第2版(英文)	2016－01	198.00	568
量子力学与经典力学之间的联系在原子、分子及电动力学系统建模中的应用(英文)	2016－01	58.00	569
算术域(英文)	2018－01	158.00	821
高等数学竞赛:1962—1991 年的米洛克斯·史怀哲竞赛(英文)	2018－01	128.00	822
用数学奥林匹克精神解决数论问题(英文)	2018－01	108.00	823
代数几何(德文)	2018－04	68.00	824
丢番图逼近论(英文)	2018－01	78.00	825
代数几何学基础教程(英文)	2018－01	98.00	826
解析数论入门课程(英文)	2018－01	78.00	827
数论中的丢番图问题(英文)	2018－01	78.00	829
数论(梦幻之旅):第五届中日数论研讨会演讲集(英文)	2018－01	68.00	830
数论新应用(英文)	2018－01	68.00	831
数论(英文)	2018－01	78.00	832

刘培杰数学工作室

已出版(即将出版)图书目录——原版影印

书　名	出版时间	定价	编号
湍流十讲(英文)	2018—04	108.00	886
无穷维李代数:第3版(英文)	2018—04	98.00	887
等值、不变量和对称性(英文)	2018—04	78.00	888
解析数论(英文)	2018—09	78.00	889
《数学原理》的演化:伯特兰·罗素撰写第二版时的 手稿与笔记(英文)	2018—04	108.00	890
哈密尔顿数学论文集(第4卷):几何学、分析学、天文学、 概率和有限差分等(英文)	2019—05	108.00	891
偏微分方程全局吸引子的特性(英文)	2018—09	108.00	979
整函数与下调和函数(英文)	2018—09	118.00	980
幂等分析(英文)	2018—09	118.00	981
李群,离散子群与不变量理论(英文)	2018—09	108.00	982
动力系统与统计力学(英文)	2018—09	118.00	983
表示论与动力系统(英文)	2018—09	118.00	984
分析学练习.第1部分(英文)	2021—01	88.00	1247
分析学练习.第2部分,非线性分析(英文)	2021—01	88.00	1248
初级统计学:循序渐进的方法:第10版(英文)	2019—05	68.00	1067
工程师与科学家微分方程用书:第4版(英文)	2019—07	58.00	1068
大学代数与三角学(英文)	2019—06	78.00	1069
培养数学能力的途径(英文)	2019—07	38.00	1070
工程师与科学家统计学:第4版(英文)	2019—06	58.00	1071
贸易与经济中的应用统计学:第6版(英文)	2019—06	58.00	1072
傅立叶级数和边值问题:第8版(英文)	2019—05	48.00	1073
通往天文学的途径:第5版(英文)	2019—05	58.00	1074
拉马努金笔记.第1卷(英文)	2019—06	165.00	1078
拉马努金笔记.第2卷(英文)	2019—06	165.00	1079
拉马努金笔记.第3卷(英文)	2019—06	165.00	1080
拉马努金笔记.第4卷(英文)	2019—06	165.00	1081
拉马努金笔记.第5卷(英文)	2019—06	165.00	1082
拉马努金遗失笔记.第1卷(英文)	2019—06	109.00	1083
拉马努金遗失笔记.第2卷(英文)	2019—06	109.00	1084
拉马努金遗失笔记.第3卷(英文)	2019—06	109.00	1085
拉马努金遗失笔记.第4卷(英文)	2019—06	109.00	1086
数论:1976年纽约洛克菲勒大学数论会议记录(英文)	2020—06	68.00	1145
数论:卡本代尔 1979:1979年在南伊利诺伊卡本代尔大学 举行的数论会议记录(英文)	2020—06	78.00	1146
数论:诺德韦克豪特 1983:1983年在诺德韦克豪特举行的 Journees Arithmetiques 数论大会会议记录(英文)	2020—06	68.00	1147
数论:1985—1988年在纽约城市大学研究生院和大学中心 举办的研讨会(英文)	2020—06	68.00	1148

刘培杰数学工作室
已出版(即将出版)图书目录——原版影印

书　名	出版时间	定　价	编号
数论:1987 年在乌尔姆举行的 Journees Arithmetiques 数论大会会议记录(英文)	2020—06	68.00	1149
数论:马德拉斯 1987:1987 年在马德拉斯安娜大学举行的国际拉马努金百年纪念大会会议记录(英文)	2020—06	68.00	1150
解析数论:1988 年在东京举行的日法研讨会会议记录(英文)	2020—06	68.00	1151
解析数论:2002 年在意大利切特拉罗举行的 C. I. M. E. 暑期班演讲集(英文)	2020—06	68.00	1152
量子世界中的蝴蝶:最迷人的量子分形故事(英文)	2020—06	118.00	1157
走进量子力学(英文)	2020—06	118.00	1158
计算物理学概论(英文)	2020—06	48.00	1159
物质,空间和时间的理论:量子理论(英文)	2020—10	48.00	1160
物质,空间和时间的理论:经典理论(英文)	2020—10	48.00	1161
量子场理论:解释世界的神秘背景(英文)	2020—07	38.00	1162
计算物理学概论(英文)	2020—06	48.00	1163
行星状星云(英文)	2020—10	38.00	1164
基本宇宙学:从亚里士多德的宇宙到大爆炸(英文)	2020—08	58.00	1165
数学磁流体力学(英文)	2020—07	58.00	1166
计算科学:第 1 卷,计算的科学(日文)	2020—07	88.00	1167
计算科学:第 2 卷,计算与宇宙(日文)	2020—07	88.00	1168
计算科学:第 3 卷,计算与物质(日文)	2020—07	88.00	1169
计算科学:第 4 卷,计算与生命(日文)	2020—07	88.00	1170
计算科学:第 5 卷,计算与地球环境(日文)	2020—07	88.00	1171
计算科学:第 6 卷,计算与社会(日文)	2020—07	88.00	1172
计算科学.别卷,超级计算机(日文)	2020—07	88.00	1173
多复变函数论(日文)	2022—06	78.00	1518
复变函数入门(日文)	2022—06	78.00	1523
代数与数论:综合方法(英文)	2020—10	78.00	1185
复分析:现代函数理论第一课(英文)	2020—07	58.00	1186
斐波那契数列和卡特兰数:导论(英文)	2020—10	68.00	1187
组合推理:计数艺术介绍(英文)	2020—07	88.00	1188
二次互反律的傅里叶分析证明(英文)	2020—07	48.00	1189
旋瓦兹分布的希尔伯特变换与应用(英文)	2020—07	58.00	1190
泛函分析:巴拿赫空间理论入门(英文)	2020—07	48.00	1191
卡塔兰数入门(英文)	2019—05	68.00	1060
测度与积分(英文)	2019—04	68.00	1059
组合学手册.第一卷(英文)	2020—06	128.00	1153
* 一代数、局部紧群和巴拿赫 * 一代数丛的表示.第一卷,群和代数的基本表示理论(英文)	2020—05	148.00	1154
电磁理论(英文)	2020—08	48.00	1193
连续介质力学中的非线性问题(英文)	2020—09	78.00	1195
多变量数学入门(英文)	2021—05	68.00	1317
偏微分方程入门(英文)	2021—05	88.00	1318
若尔当典范性:理论与实践(英文)	2021—07	68.00	1366
伽罗瓦理论.第 4 版(英文)	2021—08	88.00	1408

刘培杰数学工作室
已出版(即将出版)图书目录——原版影印

书　名	出版时间	定　价	编号
典型群,错排与素数(英文)	2020—11	58.00	1204
李代数的表示:通过 gln 进行介绍(英文)	2020—10	38.00	1205
实分析演讲集(英文)	2020—10	38.00	1206
现代分析及其应用的课程(英文)	2020—10	58.00	1207
运动中的抛射物数学(英文)	2020—10	38.00	1208
2—纽结与它们的群(英文)	2020—10	38.00	1209
概率,策略和选择:博弈与选举中的数学(英文)	2020—11	58.00	1210
分析学引论(英文)	2020—11	58.00	1211
量子群:通往流代数的路径(英文)	2020—11	38.00	1212
集合论入门(英文)	2020—10	48.00	1213
酉反射群(英文)	2020—11	58.00	1214
探索数学:吸引人的证明方式(英文)	2020—11	58.00	1215
微分拓扑短期课程(英文)	2020—10	48.00	1216
抽象凸分析(英文)	2020—11	68.00	1222
费马大定理笔记(英文)	2021—03	48.00	1223
高斯与雅可比和(英文)	2021—03	78.00	1224
π与算术几何平均:关于解析数论和计算复杂性的研究(英文)	2021—01	58.00	1225
复分析入门(英文)	2021—03	48.00	1226
爱德华·卢卡斯与素性测定(英文)	2021—03	78.00	1227
通往凸分析及其应用的简单路径(英文)	2021—01	68.00	1229
微分几何的各个方面.第一卷(英文)	2021—01	58.00	1230
微分几何的各个方面.第二卷(英文)	2020—12	58.00	1231
微分几何的各个方面.第三卷(英文)	2020—12	58.00	1232
沃克流形几何学(英文)	2020—11	58.00	1233
彷射和韦尔几何应用(英文)	2020—12	58.00	1234
双曲几何学的旋转向量空间方法(英文)	2021—02	58.00	1235
积分:分析学的关键(英文)	2020—12	48.00	1236
为有天分的新生准备的分析学基础教材(英文)	2020—11	48.00	1237
数学不等式.第一卷.对称多项式不等式(英文)	2021—03	108.00	1273
数学不等式.第二卷.对称有理不等式与对称无理不等式(英文)	2021—03	108.00	1274
数学不等式.第三卷.循环不等式与非循环不等式(英文)	2021—03	108.00	1275
数学不等式.第四卷.Jensen不等式的扩展与加细(英文)	2021—03	108.00	1276
数学不等式.第五卷.创建不等式与解不等式的其他方法(英文)	2021—04	108.00	1277

刘培杰数学工作室
已出版(即将出版)图书目录——原版影印

书　名	出版时间	定　价	编号
冯·诺依曼代数中的谱位移函数:半有限冯·诺依曼代数中的谱位移函数与谱流(英文)	2021—06	98.00	1308
链接结构:关于嵌入完全图的直线中链接单形的组合结构(英文)	2021—05	58.00	1309
代数几何方法.第1卷(英文)	2021—06	68.00	1310
代数几何方法.第2卷(英文)	2021—06	68.00	1311
代数几何方法.第3卷(英文)	2021—06	58.00	1312
代数、生物信息和机器人技术的算法问题.第四卷,独立恒等式系统(俄文)	2020—08	118.00	1199
代数、生物信息和机器人技术的算法问题.第五卷,相对覆盖性和独立可拆分恒等式系统(俄文)	2020—08	118.00	1200
代数、生物信息和机器人技术的算法问题.第六卷,恒等式和准恒等式的相等 问题、可推导性和可实现性(俄文)	2020—08	128.00	1201
分数阶微积分的应用:非局部动态过程,分数阶导热系数(俄文)	2021—01	68.00	1241
泛函分析问题与练习:第2版(俄文)	2021—01	98.00	1242
集合论、数学逻辑和算法论问题:第5版(俄文)	2021—01	98.00	1243
微分几何和拓扑短期课程(俄文)	2021—01	98.00	1244
素数规律(俄文)	2021—01	88.00	1245
无穷边值问题解的递减:无界域中的拟线性椭圆和抛物方程(俄文)	2021—01	48.00	1246
微分几何讲义(俄文)	2020—12	98.00	1253
二次型和矩阵(俄文)	2021—01	98.00	1255
积分和级数.第2卷,特殊函数(俄文)	2021—01	168.00	1258
积分和级数.第3卷,特殊函数补充:第2版(俄文)	2021—01	178.00	1264
几何图上的微分方程(俄文)	2021—01	138.00	1259
数论教程:第2版(俄文)	2021—01	98.00	1260
非阿基米德分析及其应用(俄文)	2021—03	98.00	1261
古典群和量子群的压缩(俄文)	2021—03	98.00	1263
数学分析习题集.第3卷,多元函数:第3版(俄文)	2021—03	98.00	1266
数学习题:乌拉尔国立大学数学力学系大学生奥林匹克(俄文)	2021—03	98.00	1267
柯西定理和微分方程的特解(俄文)	2021—03	98.00	1268
组合极值问题及其应用:第3版(俄文)	2021—03	98.00	1269
数学词典(俄文)	2021—01	98.00	1271
确定性混沌分析模型(俄文)	2021—06	168.00	1307
精选初等数学习题和定理.立体几何.第3版(俄文)	2021—03	68.00	1316
微分几何习题:第3版(俄文)	2021—05	98.00	1336
精选初等数学习题和定理.平面几何.第4版(俄文)	2021—05	68.00	1335
曲面理论在欧氏空间 E_n 中的直接表示(俄文)	2022—01	68.00	1444
维纳—霍普夫离散算子和托普利兹算子:某些可数赋范空间中的诺特性和可逆性(俄文)	2022—03	108.00	1496
Maple 中的数论:数论中的计算机计算(俄文)	2022—03	88.00	1497
贝尔曼和克努特问题及其概括:加法运算的复杂性(俄文)	2022—03	138.00	1498

刘培杰数学工作室
已出版(即将出版)图书目录——原版影印

书　名	出版时间	定　价	编号
复分析:共形映射(俄文)	2022—07	48.00	1542
微积分代数样条和多项式及其在数值方法中的应用(俄文)	2022—08	128.00	1543
蒙特卡罗方法中的随机过程和场模型:算法和应用(俄文)	2022—08	88.00	1544
线性椭圆型方程组:论二阶椭圆型方程的迪利克雷问题(俄文)	2022—08	98.00	1561
动态系统解的增长特性:估值、稳定性、应用(俄文)	2022—08	118.00	1565
群的自由积分解:建立和应用(俄文)	2022—08	78.00	1570
狭义相对论与广义相对论:时空与引力导论(英文)	2021—07	88.00	1319
束流物理学和粒子加速器的实践介绍:第2版(英文)	2021—07	88.00	1320
凝聚态物理中的拓扑和微分几何简介(英文)	2021—05	88.00	1321
混沌映射:动力学、分形学和快速涨落(英文)	2021—05	128.00	1322
广义相对论:黑洞、引力波和宇宙学介绍(英文)	2021—06	68.00	1323
现代分析电磁均质化(英文)	2021—06	68.00	1324
为科学家提供的基本流体动力学(英文)	2021—06	88.00	1325
视觉天文学:理解夜空的指南(英文)	2021—06	68.00	1326
物理学中的计算方法(英文)	2021—06	68.00	1327
单星的结构与演化:导论(英文)	2021—06	108.00	1328
超越居里:1903年至1963年物理界四位女性及其著名发现(英文)	2021—06	68.00	1329
范德瓦尔斯流体热力学的进展(英文)	2021—06	68.00	1330
先进的托卡马克稳定性理论(英文)	2021—06	88.00	1331
经典场论导论:基本相互作用的过程(英文)	2021—07	88.00	1332
光致电离量子动力学方法原理(英文)	2021—07	108.00	1333
经典域论和应力:能量张量(英文)	2021—05	88.00	1334
非线性太赫兹光谱的概念与应用(英文)	2021—06	68.00	1337
电磁学中的无穷空间并矢格林函数(英文)	2021—06	88.00	1338
物理科学基础数学.第1卷,齐次边值问题、傅里叶方法和特殊函数(英文)	2021—07	108.00	1339
离散量子力学(英文)	2021—07	68.00	1340
核磁共振的物理学和数学(英文)	2021—07	108.00	1341
分子水平的静电学(英文)	2021—08	68.00	1342
非线性波:理论、计算机模拟、实验(英文)	2021—06	108.00	1343
石墨烯光学:经典问题的电解决解决方案(英文)	2021—06	68.00	1344
超材料多元宇宙(英文)	2021—07	68.00	1345
银河系外的天体物理学(英文)	2021—07	68.00	1346
原子物理学(英文)	2021—07	68.00	1347
将光打结:将拓扑学应用于光学(英文)	2021—07	68.00	1348
电磁学:问题与解法(英文)	2021—07	88.00	1364
海浪的原理:介绍量子力学的技巧与应用(英文)	2021—07	108.00	1365
多孔介质中的流体:输运与相变(英文)	2021—07	68.00	1372
洛伦兹群的物理学(英文)	2021—08	68.00	1373
物理导论的数学方法和解决方法手册(英文)	2021—08	68.00	1374
非线性波数学物理学入门(英文)	2021—08	88.00	1376
波:基本原理和动力学(英文)	2021—07	68.00	1377
光电子量子计量学.第1卷,基础(英文)	2021—07	88.00	1383
光电子量子计量学.第2卷,应用与进展(英文)	2021—07	68.00	1384
复杂流的格子玻尔兹曼建模的工程应用(英文)	2021—08	68.00	1393

刘培杰数学工作室
已出版(即将出版)图书目录——原版影印

书　　名	出 版 时 间	定　价	编号
电偶极矩挑战(英文)	2021－08	108.00	1394
电动力学:问题与解法(英文)	2021－09	68.00	1395
自由电子激光的经典理论(英文)	2021－08	68.00	1397
曼哈顿计划——核武器物理学简介(英文)	2021－09	68.00	1401
粒子物理学(英文)	2021－09	68.00	1402
引力场中的量子信息(英文)	2021－09	128.00	1403
器件物理学的基本经典力学(英文)	2021－09	68.00	1404
等离子体物理及其空间应用导论.第1卷,基本原理和初步过程(英文)	2021－09	68.00	1405
拓扑与超弦理论焦点问题(英文)	2021－07	58.00	1349
应用数学:理论、方法与实践(英文)	2021－07	78.00	1350
非线性特征值问题:牛顿型方法与非线性瑞利函数(英文)	2021－07	58.00	1351
广义膨胀和齐性:利用齐性构造齐次系统的李雅普诺夫函数和控制律(英文)	2021－06	48.00	1352
解析数论焦点问题(英文)	2021－07	58.00	1353
随机微分方程:动态系统方法(英文)	2021－07	58.00	1354
经典力学与微分几何(英文)	2021－07	58.00	1355
负定相交形式流形上的瞬子模空间几何(英文)	2021－07	68.00	1356
广义卡塔兰轨道分析:广义卡塔兰轨道计算数字的方法(英文)	2021－07	48.00	1367
洛伦兹方法的变分:二维与三维洛伦兹方法(英文)	2021－08	38.00	1378
几何、分析和数论精编(英文)	2021－08	68.00	1380
从一个新角度看数论:通过遗传方法引入现实的概念(英文)	2021－07	58.00	1387
动力系统:短期课程(英文)	2021－08	68.00	1382
几何路径:理论与实践(英文)	2021－08	48.00	1385
论天体力学中某些问题的不可积性(英文)	2021－07	88.00	1396
广义斐波那契数列及其性质(英文)	2021－08	38.00	1386
对称函数和麦克唐纳多项式:余代数结构与Kawanaka恒等式(英文)	2021－09	38.00	1400
杰弗里·英格拉姆·泰勒科学论文集:第1卷.固体力学(英文)	2021－05	78.00	1360
杰弗里·英格拉姆·泰勒科学论文集:第2卷.气象学、海洋学和湍流(英文)	2021－05	68.00	1361
杰弗里·英格拉姆·泰勒科学论文集:第3卷.空气动力学以及落弹数和爆炸的力学(英文)	2021－05	68.00	1362
杰弗里·英格拉姆·泰勒科学论文集:第4卷.有关流体力学(英文)	2021－05	58.00	1363

刘培杰数学工作室
已出版(即将出版)图书目录——原版影印

书 名	出版时间	定 价	编号
非局域泛函演化方程:积分与分数阶(英文)	2021—08	48.00	1390
理论工作者的高等微分几何:纤维丛、射流流形和拉格朗日理论(英文)	2021—08	68.00	1391
半线性退化椭圆微分方程:局部定理与整体定理(英文)	2021—07	48.00	1392
非交换几何、规范理论和重整化:一般简介与非交换量子场论的重整化(英文)	2021—09	78.00	1406
数论论文集:拉普拉斯变换和带有数论系数的幂级数(俄文)	2021—09	48.00	1407
挠理论专题:相对极大值,单射与扩充模(英文)	2021—09	88.00	1410
强正则图与欧几里得若尔当代数:非通常关系中的启示(英文)	2021—10	48.00	1411
拉格朗日几何和哈密顿几何:力学的应用(英文)	2021—10	48.00	1412
时滞微分方程与差分方程的振动理论:二阶与三阶(英文)	2021—10	98.00	1417
卷积结构与几何函数理论:用以研究特定几何函数理论方向的分数阶微积分算子与卷积结构(英文)	2021—10	48.00	1418
经典数学物理的历史发展(英文)	2021—10	78.00	1419
扩展线性丢番图问题(英文)	2021—10	38.00	1420
一类混沌动力系统的分歧分析与控制:分歧分析与控制(英文)	2021—11	38.00	1421
伽利略空间和伪伽利略空间中一些特殊曲线的几何性质(英文)	2022—01	68.00	1422
一阶偏微分方程:哈密尔顿—雅可比理论(英文)	2021—11	48.00	1424
各向异性黎曼多面体的反问题:分段光滑的各向异性黎曼多面体反边界谱问题:唯一性(英文)	2021—11	38.00	1425
项目反应理论手册.第一卷,模型(英文)	2021—11	138.00	1431
项目反应理论手册.第二卷,统计工具(英文)	2021—11	118.00	1432
项目反应理论手册.第三卷,应用(英文)	2021—11	138.00	1433
二次无理数:经典数论入门(英文)	2022—05	138.00	1434
数,形与对称性:数论,几何和群论导论(英文)	2022—05	128.00	1435
有限域手册(英文)	2021—11	178.00	1436
计算数论(英文)	2021—11	148.00	1437
拟群与其表示简介(英文)	2021—11	88.00	1438
数论与密码学导论:第二版(英文)	2022—01	148.00	1423

刘培杰数学工作室
已出版(即将出版)图书目录——原版影印

书　　名	出版时间	定　价	编号
几何分析中的柯西变换与黎兹变换:解析调和容量和李普希兹调和容量、变化和振荡以及一致可求长性(英文)	2021—12	38.00	1465
近似不动点定理及其应用(英文)	2022—05	28.00	1466
局部域的相关内容解析:对局部域的扩展及其伽罗瓦群的研究(英文)	2022—01	38.00	1467
反问题的二进制恢复方法(英文)	2022—03	28.00	1468
对几何函数中某些类的各个方面的研究:复变量理论(英文)	2022—01	38.00	1469
覆盖、对应和非交换几何(英文)	2022—01	28.00	1470
最优控制理论中的随机线性调节器问题:随机最优线性调节器问题(英文)	2022—01	38.00	1473
正交分解法:涡流流体动力学应用的正交分解法(英文)	2022—01	38.00	1475
芬斯勒几何的某些问题(英文)	2022—03	38.00	1476
受限三体问题(英文)	2022—05	38.00	1477
利用马利亚万微积分进行 Greeks 的计算:连续过程、跳跃过程中的马利亚万微积分和金融领域中的 Greeks(英文)	2022—05	48.00	1478
经典分析和泛函分析的应用:分析学的应用(英文)	2022—03	38.00	1479
特殊芬斯勒空间的探究(英文)	2022—03	48.00	1480
某些图形的施泰纳距离的细谷多项式:细谷多项式与图的维纳指数(英文)	2022—03	38.00	1481
图论问题的遗传算法:在新鲜与模糊的环境中(英文)	2022—05	48.00	1482
多项式映射的渐近簇(英文)	2022—05	38.00	1483
一维系统中的混沌:符号动力学,映射序列,一致收敛和沙可夫斯基定理(英文)	2022—05	38.00	1509
多维边界层流动与传热分析:粘性流体流动的数学建模与分析(英文)	2022—05	38.00	1510
演绎理论物理学的原理:一种基于量子力学波函数的逐次置信估计的一般理论的提议(英文)	2022—05	38.00	1511
R^2 和 R^3 中的仿射弹性曲线:概念和方法(英文)	2022—08	38.00	1512
算术数列中除数函数的分布:基本内容、调查、方法、第二矩、新结果(英文)	2022—05	28.00	1513
抛物型狄拉克算子和薛定谔方程:不定常薛定谔方程的抛物型狄拉克算子及其应用(英文)	2022—07	28.00	1514
黎曼-希尔伯特问题与量子场论:可积重正化、戴森-施温格方程(英文)	2022—08	38.00	1515
代数结构和几何结构的形变理论(英文)	2022—08	48.00	1516
概率结构和模糊结构上的不动点:概率结构和直觉模糊度量空间的不动点定理(英文)	2022—08	38.00	1517

刘培杰数学工作室
已出版(即将出版)图书目录——原版影印

书　名	出版时间	定　价	编号
反若尔当对:简单反若尔当对的自同构(英文)	2022-07	28.00	1533
对某些黎曼－芬斯勒空间变换的研究:芬斯勒几何中的某些变换(英文)	2022-07	38.00	1534
内诣零流形映射的尼尔森数的阿诺索夫关系(英文)	即将出版		1535
与广义积分变换有关的分数次演算:对分数次演算的研究(英文)	即将出版		1536
强子的芬斯勒几何和吕拉几何(宇宙学方面):强子结构的芬斯勒几何和吕拉几何(拓扑缺陷)(英文)	2022-08	38.00	1537
一种基于混沌的非线性最优化问题:作业调度问题(英文)	即将出版		1538
广义概率论发展前景:关于趣味数学与置信函数实际应用的一些原创观点(英文)	即将出版		1539
纽结与物理学:第二版(英文)	2022-09	118.00	1547
正交多项式和 q-级数的前沿(英文)	2022-09	98.00	1548
算子理论问题集(英文)	2022-09	108.00	1549
抽象代数:群、环与域的应用导论:第二版(英文)	即将出版		1550
菲尔兹奖得主演讲集:第三版(英文)	即将出版		1551
多元实函数教程(英文)	2022-09	118.00	1552
球面空间形式群的几何学:第二版(英文)	2022-09	98.00	1566

联系地址:哈尔滨市南岗区复华四道街 10 号　哈尔滨工业大学出版社刘培杰数学工作室
网　　址:http://lpj.hit.edu.cn/
邮　　编:150006
联系电话:0451-86281378　　13904613167
E-mail:lpj1378@163.com